Electric Power
Transmission Systems

PRENTICE-HALL SERIES
IN ELECTRONIC TECHNOLOGY

Irving L. Kosow, editor

Charles M. Thomson, Joseph J. Gershon
and Joseph A. Labok, consulting editors.

Electric Power Transmission Systems

J. ROBERT EATON

Professor of Electrical Engineering
Purdue University
Lafayette, Indiana

PRENTICE-HALL, INC., *Englewood Cliffs, New Jersey*

10 9 8 7 6 5 4 3 2 1

ISBN: 0–13–247338–0

Library of Congress Catalog Card Number: 0–172065

Printed in the United States of America

PRENTICE-HALL INTERNATIONAL, INC., *London*
PRENTICE-HALL OF AUSTRALIA, PTY. LTD., *Sydney*
PRENTICE-HALL OF CANADA, LTD., *Toronto*
PRENTICE-HALL OF INDIA PRIVATE LIMITED, *New Delhi*
PRENTICE-HALL OF JAPAN, INC., *Tokyo*

To my wife,

Ruth

Contents

6

Preface

    This book is intended to give a first level presentation of the technical problems associated with the electric systems which transmit power from the generators to the loads. It should be of value to the technician who builds, maintains, and operates electric power equipment; to the student engineer who later goes on to more advanced study in preparation for design and analysis of power systems; and to those persons whose technical problems are closely associated with the electric power industry.

    It is assumed that the reader has had a basic course in electric circuit theory. His knowledge of mathematics should include elementary algebra and calculus, and phasor addition, multiplication, and division. Chapter 2 summarizes most of the relations used in the remainder of the book.

    This book has been produced under the sponsorship of the American Electric Power System. The first copy of the manuscript was prepared as a project of the Purdue Research Foundation.

    Many persons have provided help in the preparation of this book. These include Mr. William E. Irving of the American Electric Power Service Corporation, Mr. Paul E. Pfister and Mr. Ralph E. Snyder of the Indiana and Michigan Electric Company, Dr. Ralph Davis of the Purdue Research Foundation, Professors Gilbert L. Rainey, Stephan Freeman and R. Craig Hubele of Purdue University. Mrs. Constance Rost Philbrook was most helpful in the typing and revising of the original manuscript. Mrs. Virginia Huebner of Prentice-Hall, Inc. very effectively guided the transformation of the manuscript into book form.

    The author acknowledges the assistance of many organizations that made available data presented in this book.

The Indiana and Michigan Electric Company provided many photographs and circuit diagrams.

Public Service Indiana provided photographs of an *ac* generator and a damaged circuit breaker.

Westinghouse Electric Corporation provided tables of overhead line constants and data on transformer impedances.

General Electric Company, Power Systems Management Department, granted permission for publication of descriptions of solid state devices used in relay circuits.

Anaconda Wire and Cable Company supplied pictures of typical overhead and underground cables.

Some of the diagrams of circuit breakers are adaptations of more detailed diagrams published in bulletins of Allis Chalmers Manufacturing Company and General Electric Company.

J. R. Eaton

chapter 1

The Electric
Power System

1-1. REQUIREMENTS OF AN ELECTRIC SUPPLY SYSTEM

A great amount of effort is necessary to maintain an electric power supply within the requirements of the various types of customers served. Large investments are necessary, and continuing advancements in methods must be made as loads steadily increase from year to year. Some of the requirements for electric power supply are recognized by most consumers, such as proper voltage, availability of power on demand, reliability, and reasonable cost. Other characteristics, such as frequency, wave shape, and phase balance, are seldom recognized by the customer but are given constant attention by the utility power engineers.

The voltage of the power supply at the customer's service entrance must be held substantially constant. Variations in supply voltage are, from the consumer's view, detrimental in various respects. For example, below-normal voltage substantially reduces the light output from incandescent lamps. Above-normal voltage increases the light output but substantially reduces the life of the lamp. Motors operated at below-normal voltage draw abnormally high currents and may overheat, even when carrying no more than the rated horsepower load. Overvoltage on a motor may cause excessive heat loss in the iron of the motor, wasting energy and perhaps damaging the machine. Service voltages are usually specified by a nominal value and the voltage then maintained close to this value, deviating perhaps less then 5 percent above or below the nominal value. For example, in a 120-volt residential supply circuit, the voltage might normally vary between the limits of 115 and 125 volts as customer load and system conditions change throughout the day.

Power must be available to the consumer in any amount that he may require from minute to minute. For example, motors may be started or shut down, lights may be turned on or off, without advance warning to the electric power company. As electrical energy cannot be stored (except to a limited extent in storage batteries), the changing loads impose severe demands on the control equipment of any electrical power system. The operating staff must continually study load patterns to predict in advance those major load changes that follow known schedules, such as the starting and shutting down of factories at prescribed hours each day.

The demands for reliability of service increase daily as our industrial and social environment becomes more complex. Modern industry is almost totally dependent on electric power for its operation. Homes and office buildings are lighted, heated, and ventilated by electric power. In some instances loss of electric power may even pose a threat to life itself. Electric power, like everything else that is man-made, can never be absolutely reliable. Occasional interruptions to service in limited areas will continue. Interruptions to large areas remain a possibility, although such occurrences may be very infrequent. Further interconnection of electric supply systems over wide areas; continuing development of reliable automated control systems and apparatus; provision of additional reserve facilities; and further effort in developing personnel to engineer, design, construct, maintain, and operate these facilities will continue to improve the reliability of the electric power supply.

The cost of electric power is a prime consideration in the design and operation of electric power systems. Although the cost of almost all commodities has risen steadily over the past many years, the cost per kilowatt-hour of electrical energy has actually declined. This decrease in cost has been possible because of improved efficiencies of the generating stations and distribution systems. Although franchises often grant the electric power company exclusive rights for the supply of electric power to an area, there is keen competition between electric power and other forms of energy, particularly for heating and for certain heavy-load industrial processes.

The power-supply requirements just discussed are all well known to most electric power users. There are, however, other specifications to the electric power supply which are so effectively handled by the power companies that consumers are seldom aware that such requirements are of importance.

The frequency of electric power supply in the United States is almost entirely 60 hertz (formerly cycles per second). The frequency of a system is dependent entirely upon the speed at which the supply generator is rotated by its prime mover. Hence frequency control is basically a matter of speed control of the machines in the generating stations. Modern speed-control systems are very effective and hold frequency almost constant; deviations are seldom greater than 0.02 hertz.

In an ac system the voltage continually varies with time, at one instant being positive and a short time later being negative, going through 60 complete cycles of change in each second. Ideally a plot of the time change should be a sine wave, as shown in Fig. 1-1a. In poorly designed generating

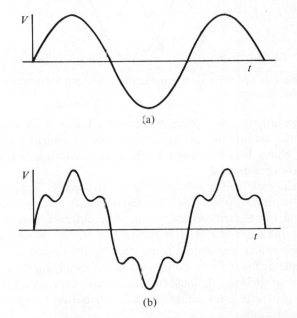

FIG. 1-1. (a) A voltage which varies sinusoidally with time. (b) A voltage wave shape which contains a fifth harmonic.

equipment, harmonics may be present and the wave shape may be somewhat as shown in Fig. 1-1b. The presence of harmonics produces unnecessary losses in the customer's equipment and sometimes produces hum in nearby telephone lines. The voltage wave shape is basically determined by the construction of the generation equipment. The power companies put specification limitations on the harmonic content of generator voltages and so require equipment manufacturers to design and build their machines to minimize trouble from this effect.

Practically all major power equipment is supplied by three-phase circuits. Three-phase circuits are essentially three single-phase circuits each of which has its own sine wave of voltage. If phase balance is perfect, each of the three voltages is of the same magnitude but displaced in time from the other two by one third of a cycle. Modern three-phase generators are designed to have almost perfect phase balance, as shown in Fig. 1-2.

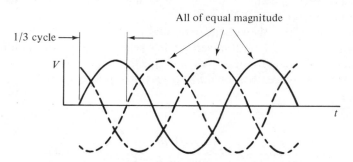

FIG. 1-2. The three voltage waves produced by a balanced three-phase generator.

Voltage unbalances produce unnecessary losses in three-phase motors and reduce the maximum starting torque. Few customers are aware of the problem of phase balance except in those rare instances where abnormal conditions are encountered.

Practically all consumers use the electric power supply to operate motors, transformers, and other devices in which iron is magnetized by currents from the electric power system. The currents that magnetize the iron do not influence the electric power bill (except where power-factor clauses apply) and so are seldom recognized by the customer. Nevertheless, provision must be made in the power system for supplying these magnetizing currents, and so their supply must be considered in specifying the characteristics of the generating, transmission, and distribution facilities that deliver power to the customer.

1-2. MINIMUM POWER SYSTEM

Electric power systems may be of great complexity and spread over large geographical areas. There is, however, a minimum system that can function as an electric power supply as shown in Fig. 1-3. The system consists of an energy source, a prime mover, a generator, and a load. Supervising it is some sort of a control system.

The *energy source* may be coal, gas, or oil burned in a furnace to heat water and generate steam in a boiler; it may be fissionable material which, in a nuclear reactor, will heat water to produce steam; it may be water in a pond at an elevation above the generating station; or it may be oil or gas burned in an internal combustion engine.

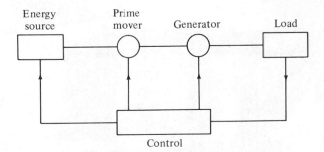

FIG. 1-3. The minimum electric power system.

FIG. 1-4. A large steam turbine which drives a generator.

The *prime mover* may be a steam-driven turbine (Fig. 1-4), a hydraulic turbine or water wheel (Fig. 1-5), or an internal combustion engine (Fig. 1-6). Each one of these prime movers has the ability to convert energy in the form of heat, falling water, or fuel into rotation of a shaft, which in turn will drive the generator.

The *generator* may be an ac machine (or alternator) (Fig. 1-7), the type of machine that supplies most of the electric power used today. In special cases the generator may be a dc machine (Fig. 1-8).

The electrical *load* on the generator in Fig. 1-3 may be lights, motors, heaters, or other devices, alone or in combination. Probably the load will vary from minute to minute as different demands occur.

The *control system* functions to keep the speed of the machines sub-

FIG. 1-5. The rotor of a 204,000 H.P. water wheel.

FIG. 1-6. A Diesel engine.

stantially constant and the voltage within prescribed limits, even though the load may change. To meet these changing load conditions, it is necessary for fuel input to change, for the prime-mover input to vary, and for the torque on the shaft from the prime mover to the generator to change

FIG. 1-7. An ac-generator rated 430,000 kVA, 24 kV, 3600 RPM.

FIG. 1-8. A dc-generator rated 6 kW, 240 volts, 1800 RPM.

7

in order that the generator may be kept at constant speed. In addition, the field current to the generator (alternator) must be adjusted to maintain constant output voltage. The control system may include a man stationed in the power plant who watches a set of meters on the generator-output terminals and makes the necessary adjustments manually. In a modern station, the control system is a *servomechanism* that senses generator-output conditions and automatically makes the necessary changes in energy input and field current to hold the electrical output within certain specifications.

1-3. MORE COMPLICATED SYSTEMS

In most situations the load is not directly connected to the generator terminals, as shown in Fig. 1-3. More commonly the load is some distance from the generator, requiring a power line connecting them, as shown in Fig. 1-9.

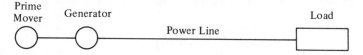

FIG. 1-9. A generator connected by a power line to a distant load.

The situation is now more complex. It is desirable to keep the electric power supply at the load within specifications. However, the controls are near the generator, which may be in another building, perhaps several miles away.

If the distance from the generator to the load is considerable, it may be desirable to install transformers at the generator and at the load end, and to transmit the power over a high-voltage line (Fig. 1-10). For the same

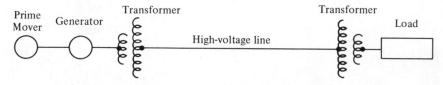

FIG. 1-10. A generator connected through transformers and a high-voltage line to a distant load.

power, the higher-voltage line carries less current, has lower losses for the same wire size, and provides more stable voltage conditions at the load. A *power transformer* is shown in Fig. 1-11, and a *high-voltage line* in Fig. 1-12. In some cases an overhead line may be be unacceptable. Instead it may be advantageous to use an *underground cable* (Fig. 1-13). With the power system shown in Fig. 1-9 or 1-10, the power supply to the load must

FIG. 1-11. A 138 kV transformer bank.

FIG. 1-12. A double-circuit 345 kV transmission line.

FIG. 1-13. Underground cables used to supply urban loads.

be interrupted if, for any reason, any component of the system must be removed from service for maintenance or repair.

Additional system load may require more power than the generator can supply. Another generator with its associated transformers and high-voltage line might be added, resulting in the system shown in Fig. 1-14.

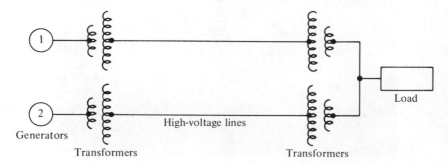

FIG. 1-14. A load supplied by two generators through separate high-voltage lines.

(*Note:* In Fig. 1-14 the prime movers of the two generators are not shown. As the discussion from now on will be confined to the electrical part of the system, the necessary prime movers will not be shown on the diagrams.)

It can be shown that there are some advantages in making ties between the generators (1) and at the ends of the high-voltage lines (2 and 3), as shown in Fig. 1-15. This system will operate satisfactorily as long as no trouble develops or no equipment needs to be taken out of service.

The above system may be vastly improved by the introduction of

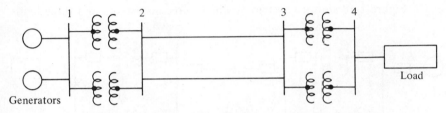

FIG. 1-15. The system of Fig. 1-14 with parallel operation of the generators, of the transformers, and of the transmission lines.

FIG. 1-16. A 345 kV oil circuit-breaker.

circuit breakers (Fig. 1-16), which may be opened and closed as needed. Circuit breakers added to the system, Fig. 1-17, permit switching out of service any selected piece of equipment without disturbing the remainder of the system. With this arrangement any element of the system may be deenergized for maintenance or repair by operation of circuit breakers. Of course, if any piece of equipment is taken out of service, the total load must then be carried by the remaining equipment. Attention must be given to avoid overloads during such circumstances. If possible, outages of equipment are scheduled at times when load requirements are below normal.

Figure 1-18 shows a system in which three generators and three loads

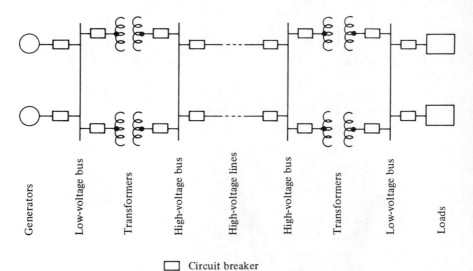

☐ Circuit breaker

FIG. 1-17. The system of Fig. 1-15 with necessary circuit-breakers.

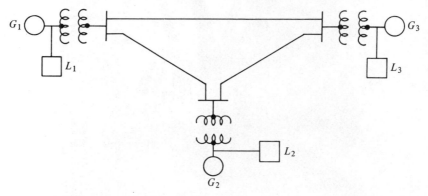

FIG. 1-18. Three generators supplying three loads over high-voltage transmission lines.

are tied together by three transmission lines. No circuit breakers are shown in this diagram, although many would be required in such a system.

1-4.   SOME TYPICAL SYSTEM LAYOUTS

The generators, lines, and other equipment which form an electric system are arranged depending on the manner in which load grows in the area and may be rearranged from time to time. Probably no two systems are exactly the same. However, there are certain plans into which a particular system design may be classified. Three types are illustrated: the radial system, the loop system, and the network system. All of these are shown without the necessary circuit breakers. In each of these systems, a single generator serves four loads.

The *radial system* is shown in Fig. 1-19. Here the lines form a "tree" spreading out from the generator. Opening any line results in interruption of power to one or more of the loads.

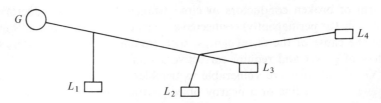

FIG. 1-19.   A radial power system supplying several loads.

The *loop system* is illustrated in Fig. 1-20. With this arrangement all loads may be served even though one line section is removed from service. In some instances during normal operation, the loop may be open at some point, such as *A*. In case a line section is to be taken out, the loop is first closed at *A* and then the line section removed. In this manner no service interruptions occur.

Figure 1-21 shows the same loads being served by a *network*. With this arrangement each load has two or more circuits over which it is fed.

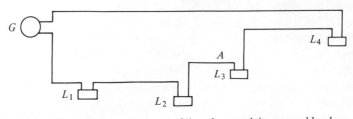

FIG. 1-20.   A loop arrangement of lines for supplying several loads.

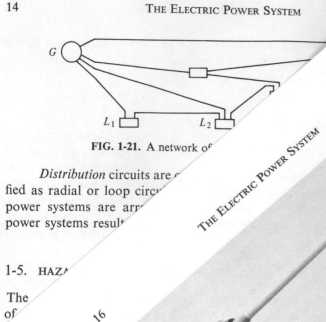

FIG. 1-21. A network o...

Distribution circuits are c...
fied as radial or loop circ...
power systems are arr...
power systems result...

1-5. HAZA...

The ...
of ...

le...
the ...

lightning ...
the line. Hi...
from conduct...
such lines are u...
and may cause no i...
directly on the power...
of such magnitude that...
an arc is established, pow...
deenergized. Lightning volta...
short duration; their time is n...
effects are most noticeable at the p...
lines, which operate at perhaps 12...
small insulators which may be flashed...
even though the lightning stroke does n...
arcs interfere with operation and usually ar...
line. When the line is deenergized, the arc is...
monly, the line may be reenergized immediately.

Overhead lines are subjected to wind and slee...
withstand ordinary storm loading, it is practically imp...
the extreme condition of sleet or wind loading. As a res...

FIG. 1-22. A 138 KV disconnect switch. The horizon...
be swung upward, hinging at the left support.

When a fault occurs on a system, conditions...
a sudden change. Voltages usually drop and currents...
are most noticeable in the immediate vicinity of...
computers, commonly called *relays* (Fig. 1-23)...
conditions, make a determination of which bre...
clear the fault, and energize the trip circuits of...
With modern equipment, the relay action an...
moval of the fault within three or four cycle...

The instruments that show circuit cor...
tect the circuits are not mounted directly...
on *switchboards* in *control houses*. Instru...
the high-voltage equipment, by means o...
the meters and relays representative sar...
ing equipment. The primary of a pote...
to the high-voltage equipment. The...
and relays a voltage which is a c...
operating equipment and is in phas...
is connected with its primary i...
winding provides a current which...
current and is in phase with it...
1-24 and 1-25.

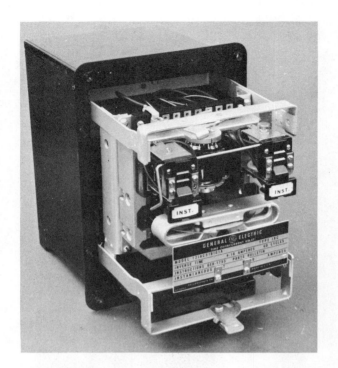

FIG. 1-23. A relay for detecting currents of excessive value.

FIG. 1-24. A 24 kV potential transformer.

FIG. 1-25. A current transformer on a 138 kV circuit.

Bushing potential devices and capacitor potential devices serve the same purpose as potential transformers but usually with less accuracy in regard to ratio and phase angle.

1-7. ONE-LINE AND THREE-LINE DIAGRAMS

Thus far, a power circuit has been represented by a single line on a diagram. This is called a *one-line diagram*. Actually, practically all power circuits consist of three conductors, and circuit breakers and disconnects must be installed in each conductor. The total circuit of a three-phase transformer bank, a three-phase circuit breaker, three-blade disconnect switches, and a three-conductor transmission line is shown in Fig. 1-26a. The one-line diagram is shown in Fig. 1-26b. One-line diagrams are used very extensively, the reader realizing that each device shown is really installed in triplicate.

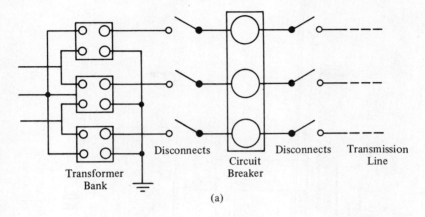

Disconnects   Disconnects   Transmission
           Circuit   Line
           Breaker

Transformer
Bank

(a)

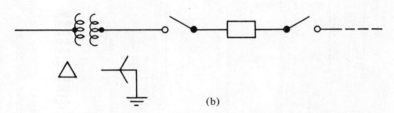

(b)

FIG. 1-26. (a) A three-phase transformer bank, the disconnect switches, and a 3-pole circuit-breaker connecting to a 3-phase transmission line. (b) A one line diagram representing the system of Fig. 1-26(a).

1-8. SOME TYPICAL POWER-SYSTEM DIAGRAMS

The layout of the major power equipment in a station is shown in Fig. 1-27. On this diagram is represented the power-system equipment from the generator terminals to the power lines that carry the electrical energy away from the plant.

A diagram of the Indiana and Michigan Electric Co. power system is shown in Fig. 1-28. This diagram shows only the generating and switching stations and the lines joining them. No details of the equipment within the stations are shown.

Practically all the electric power systems of the United States lying east of the Rocky Mountains are interconnected. Similarly, those west of the Rocky Mountains are interconnected. A section of the enormous interconnected power systems of the United States is shown in Fig. 1-29.

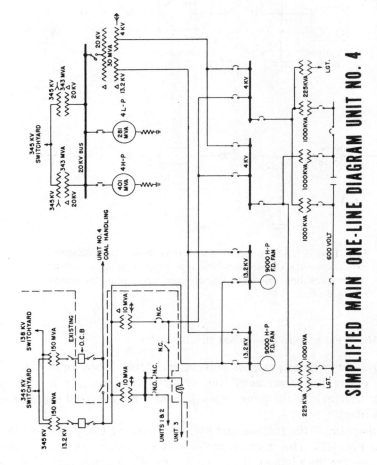

SIMPLIFIED MAIN ONE-LINE DIAGRAM UNIT NO. 4

FIG. 1-27. A simplified one-line diagram showing the circuits necessary for handling the output of a single generator and the main auxiliary equipment required for its operation.

20

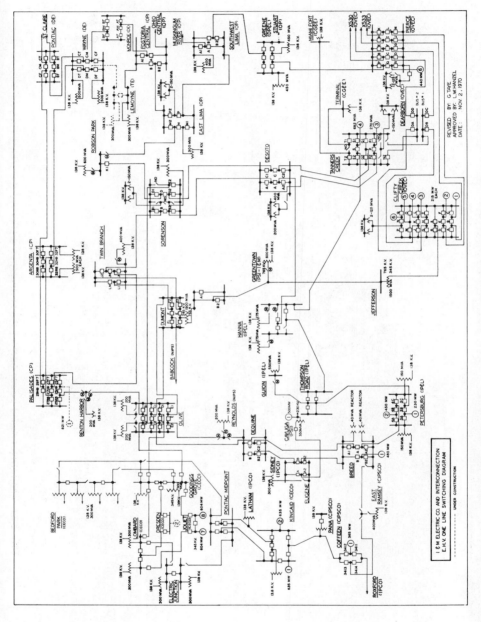

FIG. 1-28. The principal lines and interconnections of the Indiana and Michigan Electric Company.

21

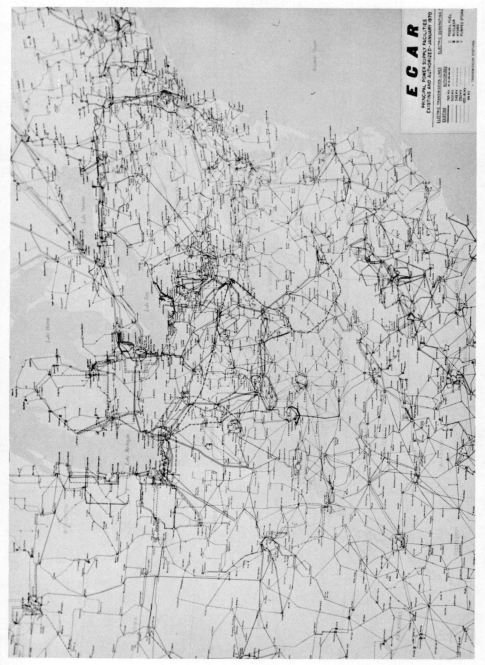

FIG. 1-29. The interconnected power lines of the northeastern section of the United States.

22

1-9. COMMUNICATION SYSTEMS

The operation of an extensive power system demands effective communications between the many stations. As a consequence, every power system also operates a communications system. The communication circuits are used for voice transmission, for the remote operation of switches and other control equipment, for the transmission of meter readings to central supervisory offices, and for relaying.

In some instances the electric power companies have constructed and maintained their own wire circuits. In other cases they may lease lines from the local telephone company. The power lines themselves may be used as communication channels by superimposing on them voltages of high frequency (50 to 100 kilohertz), which is modulated in such a fashion as to carry the intelligence desired. Microwave channels (Fig. 1-30), operating at frequencies from 2 to 6 gigahertz, may carry several hundred communication channels, involving voice, telemetering, computer control, and other functions.

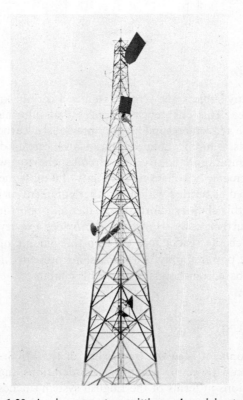

FIG. 1-30. A micro-wave transmitting and receiving tower.

chapter 2

Fundamentals of Electric Circuits

2-1. PURPOSE

In the study of any subject, the writer assumes a certain knowledge on the part of the reader. This chapter summarizes that information on electric circuits which the reader presumably has previously learned. It includes a discussion of units, symbols, notation, basic laws, certain derived relations, graphical representation, and electric circuit behavior which usually is covered in an elementary course in electricity. In subsequent chapters this information is used as needed without further comment or justification.

A study of this chapter affords the reader an opportunity to assess his preparation for further study. Hopefully, it provides a review of those basics of electricity which the reader may need in the pursuit of a study of the subject of electric power supply. All discussions presented in the remainder of this book are based on the material of this chapter.

2-2. SYSTEM OF UNITS

All subject development is in the framework of the MKS system of units. In certain instances, however, practical considerations suggest the presentation of calculations in terms of the English system (inches, pounds, etc.).

2-3. NOTATION

Table 2-1 shows a list of *physical quantities*, the units in which they are expressed, and the designating symbol. It may be noted that a few of the letter symbols are used to represent more than one quantity (such as use of L for length and also for inductance). This notation is in accordance with common practice. The relations in which these symbols are used are stated in such a fashion that no misunderstanding results.

Table 2-1. *Symbols, Units and Abbreviations*

Quantity	Symbol	Unit (MKS system)	Abbreviation
1. MASS	M	kilogram	kg
2. LENGTH	l	meter	m
3. TIME	t	second	s
4. FORCE	F	newton	N
5. CHARGE	Q	coulomb	C
6. CURRENT	I, i	ampere	A
7. ELECTROMOTIVE FORCE	E, e	volt	V
8. POTENTIAL DIFFERENCE	V, v	volt	V
9. ENERGY	W	joule	J
10. POWER	P, p	watt	W
11. RESISTANCE	R	ohm	Ω
12. CONDUCTANCE	G	mho	mho
13. RESISTIVITY	ρ	ohm-meter	Ω-m
14. CONDUCTIVITY	σ	mho per meter	mho/m
15. TEMPERATURE COEFFICIENT OF RESISTANCE	α	per degree	$/°$
16. MAGNETIC FLUX	Φ	weber	Wb
17. MAGNETIC FLUX DENSITY	B	webers per square meter	Wb/m^2
18. MAGNETOMOTIVE FORCE	$\mathscr{F}$	ampere-turn	At
19. MAGNETIC INTENSITY	H	ampere-turns per meter	At/m
20. RELUCTANCE	$\mathscr{R}$	ampere-turns per weber	At/Wb
21. PERMEANCE	$\mathscr{P}$	webers per ampere-turn	Wb/At
22. PERMEABILITY	μ	henrys per meter	H/m
23. SELF-INDUCTANCE	L	henry	H
24. MUTUAL INDUCTANCE	M	henry	H
25. ELECTRIC FLUX	ψ	line	
26. ELECTRIC FLUX DENSITY	D	lines per square meter	
27. ELECTRIC FIELD INTENSITY	$\mathscr{E}$	volts per meter	V/m
28. PERMITTIVITY	ϵ	farads per meter	F/m
29. CAPACITANCE	C	farad	F
30. ELASTANCE	S	daraf	
31. FREQUENCY	f	hertz (cycles per second)	Hz
32. ANGULAR VELOCITY	ω	radians per second	rad/s
33. PHASE ANGLE	ϕ	degree, radian	$°$ rad
34. INDUCTIVE REACTANCE	X_L	ohm	Ω

Quantity	Symbol	Unit (MKS system)	Abbreviation
35. CAPACITIVE REACTANCE	X_C	ohm	Ω
36. IMPEDANCE	Z	ohm	Ω
37. INDUCTIVE SUSCEPTANCE	B_L	mho	mho
38. CAPACITIVE SUSCEPTANCE	B_C	mho	mho
39. ADMITTANCE	Y	mho	mho
40. REACTIVE POWER	P_q	reactive volt ampere	VAR
41. APPARENT POWER	P_s	volt-ampere	VA
42. POWER FACTOR	F_p		
43. REACTIVE FACTOR	F_q		

It may be noted in Table 2-1 that voltage, current, and power are represented by the letter symbols V, I, and P, each presented in both upper-case and lowercase letters. Uppercase letters represent voltage, current, and power where all are constant, as in dc circuits. In ac circuit work, upper-case V and I represent effective values, whereas uppercase P represents average values. V_m and I_m represent the *crest*, or *maximum*, *values* of si-nusoidally varying voltage and current.

Lowercase v, i, and p represent instantaneous values of voltage, current, and power when these quantities are varying with time. Hence they may be used as *instantaneous values* of these quantities in an ac circuit.

Where needed, *double-subscript notation* is used in the representation of current and voltage. V_{AB} represents the voltage of point A with respect to point B. I_{CD} represents the current flowing through a circuit element from C to D. Note that in the circuit of Fig. 2-1, the voltage V_{AB} gives rise to the current I_{AB}.

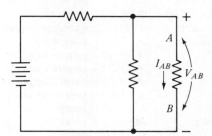

FIG. 2-1. A circuit element illustrating the polarity of the voltage and the direction of the current flow.

2-4. ELECTRIC-CIRCUIT RELATIONS

There are many well-known relations of electric circuits, some of which are regarded as basic, others are derived. The division between the two classes depends somewhat on the point of view. It is the intention here to list those relations of general importance, regardless of the classification into which they might fall.

a. General Circuit Relations. The charge Q passing a point P (Fig. 2-2) is

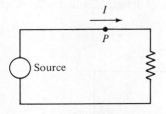

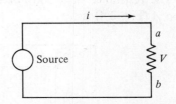

FIG. 2-2. A current I flowing past the point P.

FIG. 2-3. A circuit element carrying the current i in the presence of the voltage V.

$$Q = I t \qquad (2\text{-}1)$$

provided that the current is constant over time, t. If the current is time-varying,

$$Q = \int i \, dt \qquad (2\text{-}2)$$

from which it follows that

$$i = \frac{dQ}{dt} \qquad (2\text{-}3)$$

The power input p to a circuit element ab (Fig. 2-3) is

$$p = vi \qquad (2\text{-}4)$$

where v is the voltage across the element and i is the current through it.

The energy W delivered to the circuit element is

$$W = Pt \qquad (2\text{-}5)$$

if the power P is constant over time, t. If the power is time-varying, then

$$W = \int p \, dt \qquad (2\text{-}6)$$

b. Relations Applying to Resistors. The voltage v applied across a resistor is

$$v = iR \qquad (2\text{-}7)$$

where i is the current through the resistor and R the value of the resistance. The power p delivered to the resistor is

$$p = vi = i^2 R = \frac{v^2}{R} \tag{2-8}$$

The resistance of a body of material of length l and uniform cross section A (Fig. 2-4) is

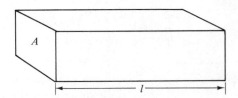

FIG. 2-4. A bar of resistance material of cross section A, length l, and resistivity ρ.

$$R = \rho \frac{l}{A} \tag{2-9}$$

where ρ is the resistivity of the material.

If the resistance of a resistor at temperature T_1 is R_1, then its resistance R_2 at temperature T_2 is

$$R_2 = R_1[1 + \alpha(T_2 - T_1)] \tag{2-10}$$

where α is the temperature coefficient of resistance at T_1.

When several resistors, R_1, R_2, and R_3 (Fig. 2-5), are connected in *series*, their equivalent resistance is

$$R_{eq} = R_1 + R_2 + R_3 \tag{2-11}$$

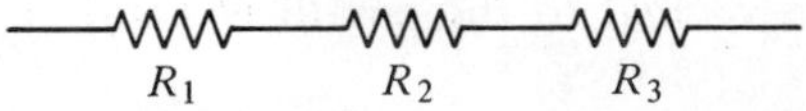

$$\text{---}\Lambda\Lambda\Lambda\text{---}\Lambda\Lambda\Lambda\text{---}\Lambda\Lambda\Lambda\text{---}$$
$$R_1 \qquad\quad R_2 \qquad\quad R_3$$

FIG. 2-5. Three resistors in simple series arrangement.

The conductance G of a resistor is

$$G = \frac{1}{R} \tag{2-12}$$

When several resistors are connected parallel (Fig. 2-6), their equivalent conductance is the sum of the individual conductances

$$G_{eq} = G_1 + G_2 + G_3 \tag{2-13}$$

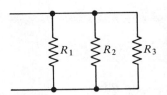

FIG. 2-6. Three resistors in parallel arrangement.

The equivalent resistance may be determined from the relation

$$\frac{1}{R_{eq}} = \frac{1}{R_1} + \frac{1}{R_2} + \frac{1}{R_3} \tag{2-14}$$

c. Relations Applying to Magnetic Devices. The magnetomotive force $\mathscr{F}$ produced by a coil of N turns carrying a current I (Fig. 2-7) is

$$\mathscr{F} = NI \tag{2-15}$$

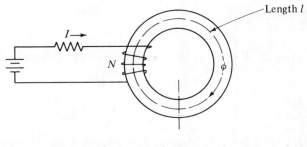

FIG. 2-7. A ring of magnetic material of length l, cross-section A, and permeability μ on which is wound a coil of N turns carrying a current I.

The magnetic flux Φ that will be produced is

$$\Phi = \frac{\mathscr{F}}{\mathscr{R}} \tag{2-16}$$

where $\mathscr{R}$ is the reluctance of the magnetic path.

The reluctance $\mathscr{R}$ of a body of material of length l and uniform cross section A is

$$\mathscr{R} = \frac{1}{\mu}\frac{l}{A} \tag{2-17}$$

where μ is the permeability of the material. For free space, $\mu_0 = 4\pi \cdot 10^{-7}$ henrys per meter.

The permeance $\mathscr{P}$ of a magnetic path of reluctance $\mathscr{R}$ is

$$\mathscr{P} = \frac{1}{\mathscr{R}} \tag{2-18}$$

The magnetic flux density B is

$$B = \frac{\Phi}{A} \tag{2-19}$$

where Φ, the magnetic flux, is assumed to be uniform over the area A.

The magnetic field intensity H is related to the magnetomotive force $\mathscr{F}$ by

$$H = \frac{\mathscr{F}}{l} \tag{2-20}$$

The permeability μ of a magnetic material is

$$\mu = \frac{B}{H} \tag{2-21}$$

If the ratio of B to H is *not* constant, the incremental permeability μ_Δ is

$$\mu_\Delta = \frac{\Delta B}{\Delta H} \tag{2-22}$$

Suppose that a magnetic field of flux density B passes from one pole face of a magnetic material across an air gap to another similar pole face of magnetic material (Fig. 2-8). A force F is produced between the two pole faces such that

$$F = \frac{B^2 A}{2\mu_0} \tag{2-23}$$

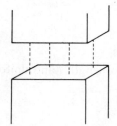

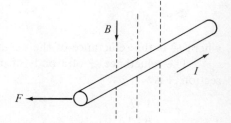

FIG. 2-8. Two magnetic poles, each of cross-section A, between which there is a magnetic field whose flux density is B.

FIG. 2-9. A conductor of length l, carrying a current I, which is at right angles to a magnetic field of flux density B.

where A is the area of the pole faces and μ_0 is the permeability of free space.

The force F produced on a conductor of length l carrying a current I in a magnetic field of density B (Fig. 2-9) is

$$F = BIl \qquad (2\text{-}24)$$

where the magnetic field is at right angles to the conductor.

Assume a magnetic flux Φ linking a coil of N turns. If the flux changes, a voltage v is generated:

$$v = N\frac{d\Phi}{dt} \qquad (2\text{-}25)$$

If the flux considered is set up by a current i in the coil itself (Fig. 2-7), then

$$\Phi = \frac{\mathcal{F}}{\mathcal{R}} = \frac{Ni}{\mathcal{R}}$$

and

$$v = \frac{NN}{\mathcal{R}}\frac{di}{dt}$$

or

$$v = L\frac{di}{dt} \qquad (2\text{-}26)$$

where the self-inductance L is

$$L = \frac{N^2}{\mathcal{R}} \qquad (2\text{-}27)$$

From Eq. (2-26),

$$i = \frac{1}{L}\int v\,dt \qquad (2\text{-}28)$$

When current I flows in an inductor of value L, the energy stored in the magnetic field is

$$W = \frac{1}{2}LI^2 \qquad (2\text{-}29)$$

Consider Eq. (2-25). If the flux Φ is set up by a current i_2 in a coil of turns N_2, which has a common magnetic circuit with coil 1 (Fig. 2-10), then

$$\Phi = \frac{\mathcal{F}_2}{\mathcal{R}} = \frac{N_2 i_2}{\mathcal{R}}$$

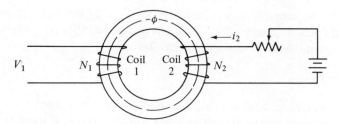

FIG. 2-10. Two coils, one with N_1 turns, the other with N_2 turns wound on the same magnetic core.

and

$$v_1 = N_1 \frac{N_2}{\mathscr{R}} \frac{di_2}{dt}$$

$$v_1 = M_{12} \frac{di_2}{dt} \tag{2-30}$$

where M_{12} is the mutual inductance between coil 1 and coil 2.

When several inductors, L_1, L_2, and L_3, are connected in series (Fig. 2-11), the equivalent inductance is

$$L_{eq} = L_1 + L_2 + L_3 \tag{2-31}$$

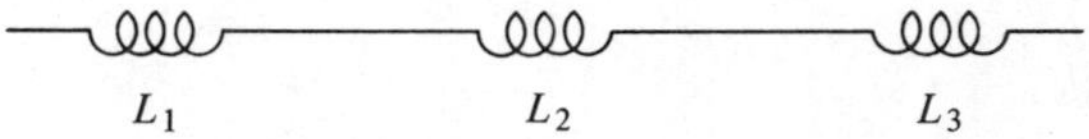

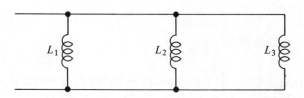

FIG. 2-11. Three inductors connected in series.

When several inductors are connected in parallel (Fig. 2-12), the equivalent inductance may be determined from the relation

FIG. 2-12. Three inductors connected in parallel.

$$\frac{1}{L_{eq}} = \frac{1}{L_1} + \frac{1}{L_2} + \frac{1}{L_3} \tag{2-32}$$

d. Relations Applying to Electric-Field Devices. If a charge Q is placed in a region of electric-field intensity $\mathscr{E}$ (Fig. 2-13), a force F is produced. Then

$$F = Q\mathscr{E} \tag{2-33}$$

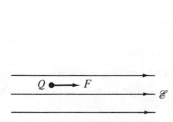

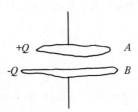

FIG. 2-13. A body carrying an electric charge Q in an electric field of intensity $\mathscr{E}$.

FIG. 2-14. A body carrying an electric charge Q from which ψ electric field lines emanate.

A body has on it a charge Q. From this body there will be ψ lines of electric flux (Fig. 2-14)

$$\psi = Q \tag{2-34}$$

The density D of the field lines will be

$$D = \frac{\psi}{A} \tag{2-35}$$

if the electric flux is uniform over the area A.

The permittivity ϵ of a dielectric material is

$$\epsilon = \frac{D}{\mathscr{E}} \tag{2-36}$$

For free space, $\epsilon_0 = 10^{-9}/36\pi$ farads per meter.

Suppose two conducting bodies A and B, separated by a dielectric, form a *capacitor* (Fig. 2-15). If one carries a charge $+Q$ and the other a

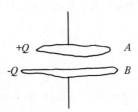

FIG. 2-15. A capacitor formed by two conducting bodies separated by a dielectric.

charge $-Q$ while the voltage between them is V, the capacitance C between the two bodies is

$$C = \frac{Q}{V} \tag{2-37}$$

The elastance S of a capacitor is

$$S = \frac{1}{C} \tag{2-38}$$

Two parallel plates each of area A and spacing d (where d is small compared to the other dimensions) form a capacitor (Fig. 2-16), whose capacitance is

$$C = \epsilon \frac{A}{d} \tag{2-39}$$

where ϵ is the permittivity of the material separating the two plates.

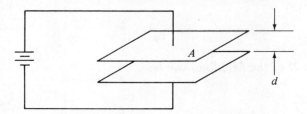

FIG. 2-16. Two parallel conducting plates of area A separated a a distance d by a dielectric of permittivity ϵ.

In a parallel-plate system as just described, the electric field intensity $\mathscr{E}$ is

$$\mathscr{E} = \frac{V}{d} \tag{2-40}$$

where V is the voltage between the two plates.

If the voltage on a capacitor is changing with time, the current required to change the charge is

$$i = \frac{dQ}{dt} = C\frac{dv}{dt} \tag{2-41}$$

From Eq. (2-41) it follows that the voltage on a capacitor is

$$v = \frac{1}{C} \int i \, dt \tag{2-42}$$

The energy W stored in the capacitor is

$$W = \tfrac{1}{2}CV^2 \tag{2-43}$$

If several capacitors, C_1, C_2, and C_3, are connected in parallel (Fig. 2-17), the equivalent capacitance is

$$C_{eq} = C_1 + C_2 + C_3 \tag{2-44}$$

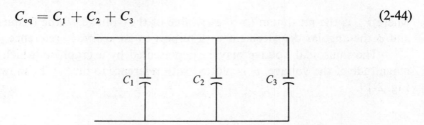

FIG. 2-17. Three capacitors connected in parallel.

If several capacitors, C_1, C_2, and C_3, are connected in series (Fig. 2-18), the equivalent elastance S_{eq} will be

$$S_{eq} = S_1 + S_2 + S_3 + \cdots \tag{2-45}$$

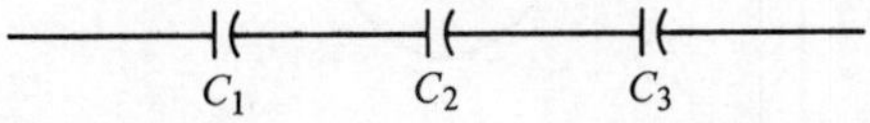

FIG. 2-18. Three capacitors connected in series.

or the equivalent capacitance may be determined from

$$\frac{1}{C_{eq}} = \frac{1}{C_1} + \frac{1}{C_2} + \frac{1}{C_3} \tag{2-46}$$

The dielectric strength (DS) of an insulating material is

$$DS = \mathscr{E}_{BD} \tag{2-47}$$

where $\mathscr{E}_{BD}$ is the electric-field intensity at which the dielectric loses its insulating property and permits ready passage of current through it, a condition know as *dielectric breakdown*.

2-5. RELATIONS PERTAINING TO AC CIRCUITS

a. Representation of Sinusoidally Varying Quantities. The currents and voltages in well-designed ac circuits vary with time following almost perfect sinusoidal patterns. A sinusoidally varying voltage (or current) may be represented in several ways. The voltage may be expressed mathematically as

$$v = V_m \cos (2\pi ft + \phi_1) \tag{2-48}$$

or

$$v = V_m \sin (2\pi ft + \phi_2) \qquad (2\text{-}49)$$

where V_m is the maximum (or crest) value of the voltage, f the frequency, and ϕ the angular displacement with respect to an arbitrary reference.

The sinusoidal voltage may be represented by a graph in which the magnitude of the voltage, v, is plotted with reference to time, t, as shown in Fig. 2-19.

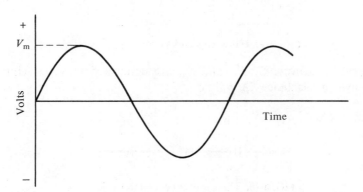

FIG. 2-19. A graph showing the voltage in an ac- circuit which varies with time as a sine wave.

A voltage may be represented by *a phasor*, as in Fig. 2-20. Here the line segment (phasor) A is assumed to have a length representative of V_m and to rotate around the point O, the angle increasing with time according to the relation

$$\phi_t = 2\pi ft + \phi_1 \qquad (2\text{-}50)$$

If we now make a plot of the horizontal projection A_h of the phasor as it varies with time, the trace which will result will conform to Eq. (2-48). If, instead, we make a plot of the vertical projection A_v of the phasor as it varies with time, the trace which will result will conform to Eq. (2-49). Present practice (followed in this textbook) favors the use of the horizontal projection and the interpretation of Eq. (2-48).

Since the phasor A has a length representative of V_m (but not necessarily equal to it), it is common practice to draw phasors expressed in terms of the effective value of the voltage rather than the maximum value:

$$V_{\text{eff}} = \frac{V_m}{\sqrt{2}} = 0.707 V_m \qquad (2\text{-}51)$$

A phasor may be expressed in several ways. We shall use the symbol $\bar{A}$ to represent a phasor of length A set at a particular angle ϕ with respect to a reference, as in Fig. 2-21. We may express A in complex notation. In rectangular form

$$\bar{A} = m + jn \tag{2-52}$$

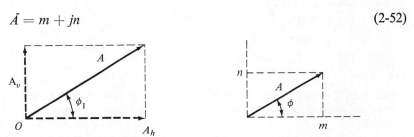

FIG. 2-20. A phasor representing a voltage which varies sinusoidally with time. The magnitude of the voltage is A.

FIG. 2-21. A phasor of length A, whose horizontal component is m, whose vertical component is n, and which makes an angle ϕ with the reference axis.

indicating that phasor A may be regarded as the sum of two phasors at right angles to each other ($j = \sqrt{-1}$). Or we may write

$$\bar{A} = A\underline{/\phi} \tag{2-53}$$

which is the *polar form* of Eq. (2-52). Or we may write

$$\bar{A} = A\epsilon^{j\phi} \tag{2-54}$$

which is the *exponential form* of Eq. (2-52).

b. Addition of Sine Waves. Two (or more) sinusoids v_1 and v_2 may be added when represented by either of the three methods of expression just described. For example,

$$v_t = V_{1m} \cos (2\pi f_1 t + \phi_1) + V_{2m} \cos (2\pi f_2 t + \phi_2) \tag{2-55}$$

If f_1 is equal to f_2, mathematical manipulation will show that v_t is a sinusoid of the same frequency:

$$v_t = V_{tm} \cos (2\pi f t + \phi_3) \tag{2-56}$$

The two waves of Eq. (2-55) are shown graphically in Fig. 2-22. These two waves may be added point by point to give v_t as shown.

The two waves of Eq. (2-55) may be represented by phasors (Fig. 2-23). By adding the phasors, following the law of vector addition, a new phasor V_t

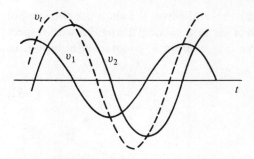
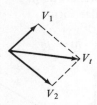

FIG. 2-22. Two sine waves (of voltage) v_1 and v_2 which when added point by point give the total v_t.

FIG. 2-23. Two phasors V_1 and V_2 representing sine waves which when added as vectors, give V_t, which in turn represents the sum.

is produced which properly represents the sum. This addition may be performed either graphically or analytically.

c. Resistance. The expression relating voltage and current in a resistor is

$$v = iR \qquad (2\text{-}7)$$

If the current through the resistor is a sinusoid represented by

$$i = I_m \cos 2\pi ft \qquad (2\text{-}57)$$

then

$$v = I_m R \cos 2\pi ft$$
$$= V_m \cos 2\pi ft \qquad (2\text{-}58)$$

where

$$V_m = I_m R$$

or

$$V = IR \qquad (2\text{-}59)$$

where V and I are effective values.

A comparison of Eqs. (2-57) and (2-58) shows that the wave of voltage is in phase with the wave of current (Fig. 2-24). On a phasor diagram, the voltage and current are represented as shown in Fig. 2-25a, where both phasors are along the line of reference, or as in Fig. 2-25b, where both are

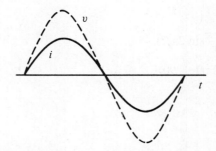

FIG. 2-24. In a resistor, the sine waves of current and voltage are in phase with each other.

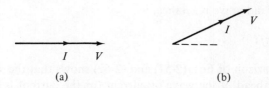

(a)   (b)

FIG. 2-25. Phasor diagrams showing the current and voltage of a resistor in phase with each other. (a) Phasors are on the reference axis. (b) Phasors are at an angle with the reference axis.

at an angle with reference. Since it is presumed that these phasors rotate with time, either position is correct. The *relative* position of the two phasors is of importance.

In phasor notation

$$\bar{V}_R = I_R R \tag{2-60}$$

The power to the resistor,

$$p = vi \tag{2-4}$$

when averaged over a complete cycle is

$$P_R = VI = I^2 R \tag{2-61}$$

where V and I are effective values.

d. Inductance. In an inductor

$$v = L \frac{di}{dt} \tag{2-26}$$

If, as in the case of the resistor, the current is represented by

$$i = I_m \cos 2\pi ft \tag{2-57}$$

then

$$v = I_m L 2\pi f(-\sin 2\pi ft) \tag{2-62a}$$

$$v = V_m \cos(2\pi ft + 90) \tag{2-62b}$$

where

$$V_m = I_m 2\pi fL \tag{2-63}$$

or

$$V_m = I_m X_L \tag{2-64}$$

where X_L, the inductive reactance, is

$$X_L = 2\pi fL \tag{2-65}$$

A comparison of Eqs. (2-57) and (2-62) shows that the wave of voltage is 90 degrees ahead of the wave of current (or the current is 90° behind the voltage), Fig. 2-26.

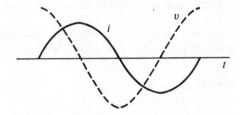

FIG. 2-26. In an inductor, the sine wave of current lags the sine wave of voltage by 90°.

On a phasor diagram, the voltage and current are represented as shown in Fig. 2-27a or b. From this diagram and Eq. (2-64) it follows that if the current and voltage are represented by phasors, then

$$\bar{V}_L = jX_L I = j2\pi fL I \tag{2-66}$$

where the j indicates a 90° displacement.

The instantaneous power to the inductor,

$$p = vi \tag{2-5}$$

when averaged over a complete cycle is

$$P_L = 0 \tag{2-67}$$

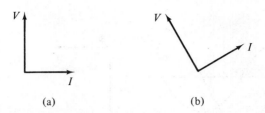

(a)   (b)

FIG. 2-27. Phasor diagrams showing the current through an inductor lagging the voltage across it by 90°. (a) The current phasor is on the reference axis. (b) The current and voltage phasors are displaced from the reference axis.

e. Capacitance. In a capacitor

$$v = \frac{1}{C} \int i \, dt \tag{2-42}$$

If, as in the case of the resistor, the current is represented by

$$i = I_m \cos 2\pi f t \tag{2-57}$$

then

$$v = I_m \frac{1}{2\pi f C} \sin 2\pi f t$$

or

$$V = V_m \cos (2\pi f t - 90°) \tag{2-68}$$

where

$$V_m = I_m \frac{1}{2\pi f C} \tag{2-69}$$

or

$$V_m = I_m X_C \tag{2-70}$$

where X_C, the capacitive reactance, is

$$X_C = \frac{1}{2\pi f C} \tag{2-71}$$

A comparison of Eqs. (2-57) and (2-68) shows that the wave of voltage is 90 degrees behind the wave of current (Fig. 2-28).

On a phasor diagram, the voltage and current are shown as in Fig. 2-29a or b. From this diagram and Eq. (2-71) it follows that if the current

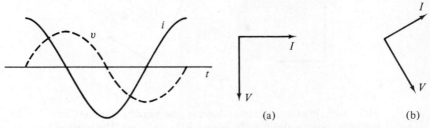

(a)   (b)

FIG. 2-28. In a capacitor, the sine wave of current leads the sine wave of voltage by 90°.

FIG. 2-29. Phasor diagrams showing the current of a capacitor leading the voltage across it by 90°. (a) The current phasor is on the reference axis. (b) The current and voltage phasors are displaced from the reference axis.

and voltage are represented by phasors, then

$$\bar{V}_C = -jX_C\bar{I}_C = \frac{-j}{2\pi f C}\bar{I}_C \tag{2-72}$$

where the $-j$ indicates a $-90°$ displacement.

The instantaneous power to the capacitor,

$$p = vi \tag{2-5}$$

when averaged over a complete cycle is

$$P_C = 0 \tag{2-73}$$

f. Series Combinations of Resistance, Inductance, and Capacitance. Consider a resistor in series with an inductor carrying a sinusoidally varying current I (Fig. 2-30a). As the current is common to both circuit elements, it will be used as reference in the phasor diagram (Fig. 2-30b). The voltage across the resistor, V_R, is in phase with I,

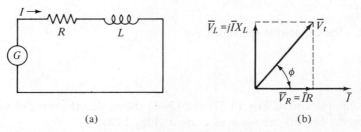

(a)   (b)

FIG. 2-30. A series combination of a resistor R and an inductor L. (a) The circuit diagram. (b) The phasor diagram.

while the voltage across the inductor, V_L, leads the current by 90°. By phasor addition, the total voltage V_t is

$$V_t = \sqrt{V_R^2 + V_L^2} \qquad\qquad (2\text{-}74)$$

$$\phi = \arctan\frac{V_L}{V_R} \qquad\qquad (2\text{-}75)$$

since (in phasor notation)

$$\bar{V}_R = IR \qquad\qquad (2\text{-}60)$$

$$\bar{V}_L = Ij2\pi fL = IjX_L \qquad\qquad (2\text{-}66)$$

$$\bar{V}_t = I(R + jX_L) \qquad\qquad (2\text{-}76)$$
$$= IZ$$

then Z, the circuit impedance, is

$$Z = R + jX_L \qquad\qquad (2\text{-}77)$$

Next, consider a resistor in series with a capacitor in an ac circuit (Fig. 2-31a). An analysis of the phasor diagrams for this circuit (Fig. 2-31b)

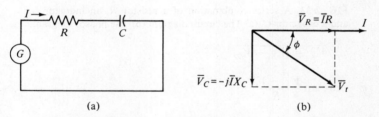

(a)   (b)

FIG. 2-31. A series combination of a resistor R and a capacitor C. (a) The circuit diagram. (b) The phasor diagram.

shows that

$$V_t = \sqrt{V_R^2 + V_C^2} \qquad\qquad (2\text{-}78)$$

$$\phi = -\arctan\frac{V_C}{V_R} \qquad\qquad (2\text{-}79)$$

$$\bar{V}_R = IR \qquad\qquad (2\text{-}60)$$

$$\bar{V}_C = -jX_C I = \frac{-j}{2\pi fC}I \qquad\qquad (2\text{-}72)$$

and

$$\bar{V}_t = I(R - jX_c)$$
$$= IZ \qquad (2\text{-}80)$$

where

$$Z = R - jX_c \qquad (2\text{-}81)$$

A similar analysis of an ac circuit consisting of a series combination of a resistor, an inductor, and a capacitor (Fig. 2-32a) and the corresponding phasor diagram (Fig. 2-32b) shows that

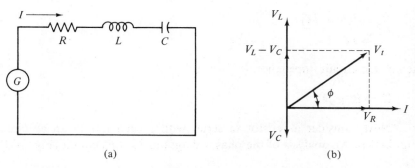

(a)   (b)

FIG. 2-32. A series combination of a resistor R, an inductor L, and a capacitor C. (a) The circuit diagram (b) The phasor diagram.

$$V_t = \sqrt{V_r^2 + (V_L - V_c)^2} \qquad (2\text{-}82)$$

$$\phi = \arctan \frac{V_L - V_c}{R} \qquad (2\text{-}83)$$

In phasor notation

$$\bar{V}_t = I\left(R + j\omega L - \frac{j}{\omega C}\right) \qquad \omega = 2\pi f \qquad (2\text{-}84)$$

$$\bar{V}_t = IZ \qquad (2\text{-}85)$$

where

$$Z = R + jX_L - jX_c \qquad (2\text{-}86)$$

An RLC series circuit (Fig. 2-32a) is said to be *resonant* when

$$X_L - X_c = 0 \qquad (2\text{-}87)$$

Then

$$Z = R \qquad (2\text{-}88)$$

Since

$$X_L = 2\pi f L$$

and

$$X_C = \frac{1}{2\pi f C}$$

it follows that a circuit may be brought to resonance by a proper adjustment of either inductance, capacitance, or frequency.

g. Parallel Combinations of Resistance, Inductance, and Capacitance. Consider an ac circuit in which a resistor, an inductor, and a capacitor are in parallel (Fig. 2-33a). Note that V is

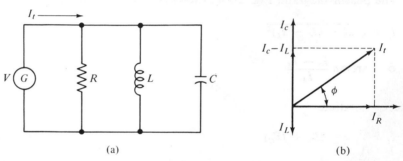

(a)   (b)

FIG. 2-33. A parallel combination of a resistor R, an inductor L, and a capacitor C. (a) The circuit diagram. (b) The phasor diagram.

common to R, L, and C and is therefore used as the reference in Fig. 2-33b. By Eq. (2-60),

$$\bar{I}_R = \frac{\bar{V}}{R}$$

or

$$\bar{I}_R = \bar{V}G \qquad (2\text{-}89)$$

where

$$G = \frac{1}{R} \qquad (2\text{-}90)$$

By Eq. (2-66),

$$I = \frac{\bar{V}}{jX_L} = \bar{V}(-jB_L)$$

where

$$B_L = \frac{1}{X_L} = \frac{1}{2\pi f L}$$

By Eq. (2-72),

$$I_c = \frac{\bar{V}}{-jX_c} = \bar{V}j\text{—}$$

where

$$B_c =$$

No...

In phasor ...

$$I_t = \bar{V}(G \cdot$$

$$I_t = \bar{V}Y$$

where

$$Y = G - jB_L + jB_c$$

A parallel circuit is said to ...

$$-B_L + B_c = 0$$

Then

$$Y = G$$

The condition of parallel resonance may be produ...
ment of either inductance, capacitance, or frequency.

The reactive power, P_q, applying to all ac circuits, is

$$P_q = VI \sin\phi \qquad \text{reactive volt-amperes} \qquad (2\text{-}106)$$

The power factor, F_p, is

$$F_p = \cos\phi = \frac{P}{P_s} \qquad (2\text{-}10$$

The reactive factor, F_q, is

$$F_q = \sin\phi = \frac{P_q}{P_s}$$

A lagging load is one in which the current lags the voltag...
inductive circuit. A leading load is one in which the current lea...
age, as in a capacitive circuit.

The reactive power supplies to an inductor is termed p...
a capacitor is termed negative.

A diagram showing a geometric relationship betwee...
shown in Fig. 2-34. Here it may be noted that

$$P_s = \sqrt{P^2 + P_q^2}$$

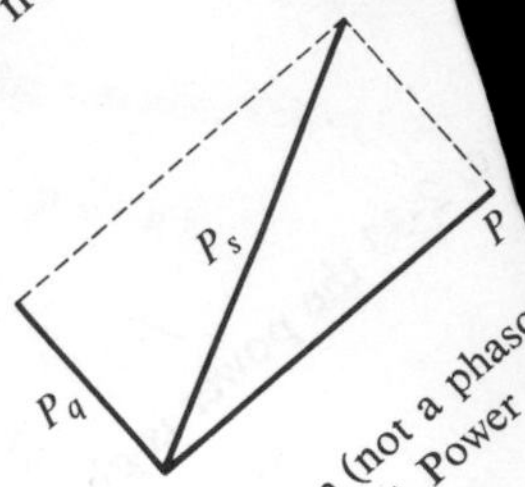

FIG. 2-34. A vector diagram (not a phasor...
relationship between Apparent Power P_s...
Reactive (or Quadrature) Power P_q.

2-6. NETWORK RELATIONS

a. Kirchhoff's Voltage ...
algebraic sum of all voltages is zero.
The law is illustrated in Fig...
path)

$$V_{1k} + V_{ab} + V_{ef} + V_{gh} =$$

The law is also illustrated by ...

Then

$$Z = R \tag{2-88}$$

Since

$$X_L = 2\pi f L$$

and

$$X_C = \frac{1}{2\pi f C}$$

it follows that a circuit may be brought to resonance by a proper adjustment of either inductance, capacitance, or frequency.

g. Parallel Combinations of Resistance, Inductance, and Capacitance. Consider an ac circuit in which a resistor, an inductor, and a capacitor are in parallel (Fig. 2-33a). Note that V is

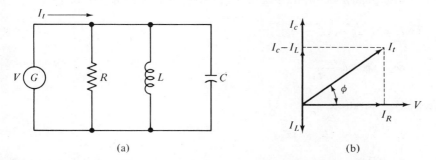

(a)   (b)

FIG. 2-33. A parallel combination of a resistor R, an inductor L, and a capacitor C. (a) The circuit diagram. (b) The phasor diagram.

common to R, L, and C and is therefore used as the reference in Fig. 2-33b. By Eq. (2-60),

$$I_R = \frac{\bar{V}}{R}$$

or

$$I_R = \bar{V}G \tag{2-89}$$

where

$$G = \frac{1}{R} \tag{2-90}$$

By Eq. (2-66),

$$I = \frac{\bar{V}}{jX_L} = \bar{V}(-jB_L) \qquad (2\text{-}91)$$

where

$$B_L = \frac{1}{X_L} = \frac{1}{2\pi f L} \qquad (2\text{-}92)$$

By Eq. (2-72),

$$I_C = \frac{\bar{V}}{-jX_C} = \bar{V}jB_C \qquad (2\text{-}93)$$

where

$$B_C = \frac{1}{X_C} = 2\pi f C \qquad (2\text{-}94)$$

Note that R, X_L, X_C, G, B_L, and B_C are all *real* numbers, not complex. The phasor diagram, Fig. 2-33b, shows the total current to be

$$I_t = \sqrt{I_R^2 + (I_C - I_L)^2} \qquad (2\text{-}95)$$

$$\phi = \arctan \frac{I_C - I_L}{I_R} \qquad (2\text{-}96)$$

In phasor notation,

$$I_t = \bar{V}(G - jB_L + jB_C) \qquad (2\text{-}97)$$

$$I_t = \bar{V}Y \qquad (2\text{-}98)$$

where

$$Y = G - jB_L + jB_C \qquad (2\text{-}99)$$

A parallel circuit is said to be resonant when

$$-B_L + B_C = 0 \qquad (2\text{-}100)$$

Then

$$Y = G \qquad (2\text{-}101)$$

The condition of parallel resonance may be produced by a proper adjustment of either inductance, capacitance, or frequency.

h. Power, Apparent Power, and Reactive Power.
From Eq. (2-61) it may be noted that in an ac circuit, the power to a resistor averaged over a cycle is

$$P_R = I^2 R \qquad (2\text{-}61)$$

where I is the effective value of the current.

From Eqs. (2-67) and (2-73), it may be noted that in an ac circuit the power to an inductor or to a capacitor averaged over a cycle is

$$P_L = 0 \qquad (2\text{-}67)$$

$$P_C = 0 \qquad (2\text{-}73)$$

As a consequence, to determine the power dissipated in an ac circuit it is necessary to look only at the power dissipated in the resistors.

In the series RLC circuit of Fig. 2-32 the power is given by

$$P = V_R I$$

However, we may write

$$V_R = V_t \cos \phi$$

or

$$P = V_t I \cos \phi \qquad (2\text{-}102)$$

In the parallel circuit of Fig. 2-33 the power is given by

$$P = V I_R$$

Since

$$I_R = I_t \cos \phi$$

$$P = V I_t \cos \phi \qquad (2\text{-}103)$$

It can be shown that in any circuit the active power is given by

$$P = V I \cos \phi \qquad \text{watts} \qquad (2\text{-}104)$$

where ϕ is the angle between the voltage V and the current I.

The vector power or apparent power, P_s, applying to all ac circuits, is

$$P_s = V I \qquad \text{volt-amperes} \qquad (2\text{-}105)$$

The reactive power, P_q, applying to all ac circuits, is

$$P_q = VI \sin \phi \qquad \text{reactive volt-amperes} \qquad (2\text{-}106)$$

The power factor, F_p, is

$$F_p = \cos \phi = \frac{P}{P_s} \qquad (2\text{-}107)$$

The reactive factor, F_q, is

$$F_q = \sin \phi = \frac{P_q}{P_s} \qquad (2\text{-}108)$$

A lagging load is one in which the current lags the voltage, as in an inductive circuit. A leading load is one in which the current leads the voltage, as in a capacitive circuit.

The reactive power supplies to an inductor is termed positive, that to a capacitor is termed negative.

A diagram showing a geometric relationship between P_s, P, and P_q is shown in Fig. 2-34. Here it may be noted that

$$P_s = \sqrt{P^2 + P_q^2} \qquad (2\text{-}109)$$

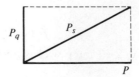

FIG. 2-34. A vector diagram (not a phasor diagram) showing the relationship between Apparent Power P_s, Active Power P, and Reactive (or Quadrature) Power P_q.

2-6. NETWORK RELATIONS

a. Kirchhoff's Voltage Law. Around any closed path, the algebraic sum of all voltages is zero.

The law is illustrated in Fig. 2-35 by writing (for the closed-circuit path)

$$V_{lk} + V_{ab} + V_{ef} + V_{gh} = 0$$

The law is also illustrated by writing

are tied together by three transmission lines. No circuit breakers are shown in this diagram, although many would be required in such a system.

1-4. SOME TYPICAL SYSTEM LAYOUTS

The generators, lines, and other equipment which form an electric system are arranged depending on the manner in which load grows in the area and may be rearranged from time to time. Probably no two systems are exactly the same. However, there are certain plans into which a particular system design may be classified. Three types are illustrated: the radial system, the loop system, and the network system. All of these are shown without the necessary circuit breakers. In each of these systems, a single generator serves four loads.

The *radial system* is shown in Fig. 1-19. Here the lines form a "tree" spreading out from the generator. Opening any line results in interruption of power to one or more of the loads.

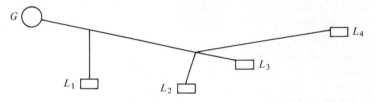

FIG. 1-19. A radial power system supplying several loads.

The *loop system* is illustrated in Fig. 1-20. With this arrangement all loads may be served even though one line section is removed from service. In some instances during normal operation, the loop may be open at some point, such as *A*. In case a line section is to be taken out, the loop is first closed at *A* and then the line section removed. In this manner no service interruptions occur.

Figure 1-21 shows the same loads being served by a *network*. With this arrangement each load has two or more circuits over which it is fed.

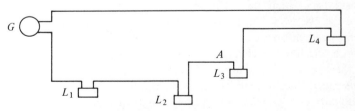

FIG. 1-20. A loop arrangement of lines for supplying several loads.

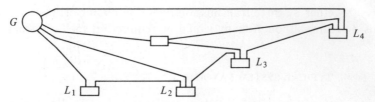

FIG. 1-21. A network of lines supplying several loads.

Distribution circuits are commonly designed so that they may be classified as radial or loop circuits. The high-voltage transmission lines of most power systems are arranged as networks. The interconnection of major power systems results in networks made up of many line sections.

1-5. HAZARDS TO POWER-SYSTEM OPERATION

The operation of electric power systems may be disturbed as the result of a number of causes. The troubles on the electric circuits are usually in the form of broken conductors or circumstances in which conductors are temporarily (or permanently) connected to each other or to ground. Regardless of the cause of the disturbances, such troubles seriously interfere with the flow of power and require corrective action.

Overhead lines are vulnerable to troubles set up by lightning. When lightning strikes a line or a nearby object, a *transient voltage* is created on the line. High-voltage circuits have large spacings between conductors and from conductors to grounded objects, such as towers. Direct strokes to such lines are usually diverted to ground through overhead ground wires and may cause no interference with operation. Occasionally strokes terminate directly on the power conductors and introduce on them transient voltages of such magnitude that *arcover* across insulator strings may result. Once an arc is established, power current flows through it until the circuit is deenergized. Lightning voltages may be of high magnitude but are of very short duration; their time is measured in microseconds. As a result, their effects are most noticeable at the point where the stroke occurs. Distribution lines, which operate at perhaps 12 kV or less, are mounted on relatively small insulators which may be flashed over by the voltage induced on a line even though the lightning stroke does not contact the line. Such resulting arcs interfere with operation and usually are eliminated by deenergizing the line. When the line is deenergized, the arc is extinguished and, very commonly, the line may be reenergized immediately.

Overhead lines are subjected to wind and sleet. Although designed to withstand ordinary storm loading, it is practically impossible to design for the extreme condition of sleet or wind loading. As a result, occasional line failures occur as a result of these causes. Failures resulting from wind or sleet are usually of such nature that repairs must be made before the line may be reenergized.

Foreign objects may contact a line and cause damage which necessitates removal of the line from service. Airplanes may fly into a line, an automobile may knock down a pole, a construction worker may run a crane into a line, or an excavator may dig up a cable. Birds, squirrels, snakes, and other animals occasionally short-circuit lines and cause outages.

Generators, transformers, lines, or other components may be switched out of service for normal or routine maintenance. When switching operations occur, transients are set up on the power system which sometimes produce serious overvoltages. Such overvoltages in turn may cause flashover of insulators or puncture of solid insulation.

Equipment sometimes fails from no particular direct cause. Failure may be due to inadequate design, to previous overloads, or to natural deterioration of equipment.

Human errors may produce power-system outages. Switches may be opened unintentionally cutting off loads; temporary grounding electrodes may be put on energized circuits by mistake; equipment handled by hot-line devices may be dropped. Such occasions, although rare, must be considered as hazard possibilities in power-system design.

1-6. NEED FOR AUXILIARY EQUIPMENT

Circuit breakers such as those shown in Fig. 1-16 are necessary to deenergize equipment either for normal operation or on the occurrence of short circuits. Circuit breakers must be designed to carry normal-load currents continuously, to withstand the extremely high currents that occur during faults, and to separate contacts and clear a circuit in the presence of a fault. Circuit breakers are rated in terms of these duties.

When a circuit breaker opens to deenergize a piece of equipment, one side of the circuit breaker usually remains energized, as it is connected to operating equipment. Since it is sometimes necessary to work on the circuit breaker itself, it is also necessary to have means by which the circuit breaker may be completely disconnected from other energized equipment. For this purpose *disconnect switches* (Fig. 1-22) are placed in series with the circuit breakers. By opening these disconnects, the circuit breaker may be completely deenergized, permitting work to be carried on in safety.

Various *instruments* are necessary to monitor the operation of the electric power system. Usually each generator, each transformer bank, and each line has its own set of instruments, frequently consisting of *voltmeters*, *ammeters*, *wattmeters*, and *varmeters*.

FIG. 1-22. A 138 kV disconnect switch. The horizontal bar may be swung upward, hinging at the left support.

When a fault occurs on a system, conditions on the system undergo a sudden change. Voltages usually drop and currents increase. These changes are most noticeable in the immediate vicinity of the fault. On-line *analog computers*, commonly called *relays* (Fig. 1-23) monitor these changes of conditions, make a determination of which breakers should be opened to clear the fault, and energize the trip circuits of those appropriate breakers. With modern equipment, the relay action and breaker opening causes removal of the fault within three or four cycles after its initiation.

The instruments that show circuit conditions and the relays that protect the circuits are not mounted directly on the power lines but are placed on *switchboards* in *control houses*. Instrument transformers are installed on the high-voltage equipment, by means of which it is possible to pass on to the meters and relays representative samples of the conditions on the operating equipment. The primary of a potential transformer is connected directly to the high-voltage equipment. The secondary provides for the instruments and relays a voltage which is a constant fraction of the voltage on the operating equipment and is in phase with it. Similarly, a current transformer is connected with its primary in the high-voltage circuit. The secondary winding provides a current which is a known fraction of the power-equipment current and is in phase with it. Instrument transformers are shown in Figs. 1-24 and 1-25.

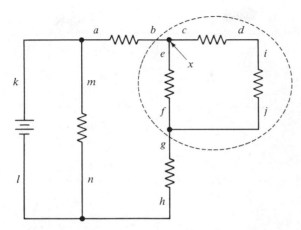

FIG. 2-35. A circuit to illustrate Kirchhoff's Voltage Law and Current Law.

$$V_{lk} + V_{ab} + V_{cd} + V_{dl} = 0$$

Here the path is closed, but it does not follow the branches of the circuit.

b. Kirchhoff's Current Law. The algebraic sum of all the currents to a point (or to a bounded region) is zero. This law is illustrated in Fig. 2-35 by writing (for point X)

$$I_{ab} + I_{dc} + I_{fe} = 0$$

It is further illustrated by writing (for the region bounded by the dashed line)

$$I_{ab} + I_{hg} = 0$$

c. Superposition Theorem. In a network containing several sources (voltage or current) the current in any branch may be determined by calculating the current in that branch in the presence of a single source while all other sources are reduced to zero, repeating the process for each source in turn, and then algebraically summing the several currents so calculated. This theorem is illustrated in Fig. 2-36. The current in branch X of (a) is the algebraic sum of the current in branch X determined in (b) plus the current in branch X determined in (c).

d. Thévenin's Theorem. In a network consisting of any number of sources and any number of branches, the behavior of a load attached to any two terminals *mn* may be analyzed by considering the network to be replaced by a simple circuit consisting of a single voltage source and a single impedance in series connection to the load.

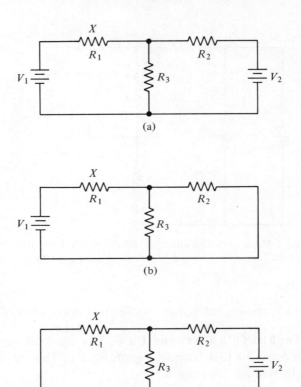

FIG. 2-36. Illustrating the Superposition Theorem. The current in Branch X of (a) is equal to the sum of the current in Branch X of (b) and the current in Branch X of (c).

The Thévenin voltage is the voltage measured between the load terminals mn on the total network but with no load on mn. The Thévenin impedance is that impedance measured between terminals mn of the network with all source voltages and source currents reduced to zero and with no load connected to mn. This theorem is illustrated in Fig. 2-37, in which (a) is the total circuit with the load connected. In (b) the Thévenin voltage is being measured. In (c) the Thévenin impedance is being measured. In (d) the Thévenin equivalent circuit is presented with the load connected.

    e. Delta–Wye and Wye–Delta Transformations. Three points A, B, and C in an electric network may sometimes be interconnected by three impedances Z_{ab}, Z_{bc}, and Z_{ca} as shown in Fig. 2-38a. Such an arrangement of impedances is said to be *delta-connected*. In studies involving circuit reduction or in three-phase-circuit analysis, it is sometimes advan-

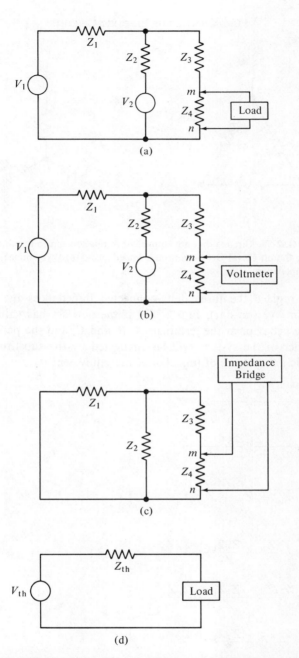

FIG. 2-37. Illustrating Thévenin's Theorem (a) A load is connected across two terminals *mn* of a complicated network. (b) The load is removed and the voltage V_{Th} between the terminals *mn* is measured. (c) The sources are removed and the impedance Z_{Th} between terminals *mn* is measured. (d) The Thévenin equivalent circuit.

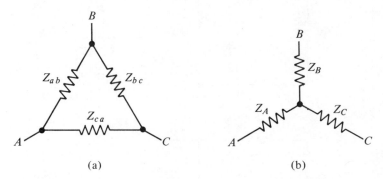

FIG. 2-38. Circuits that are equivalent in relationship to terminals *A*, *B*, and *C*. (a) Delta-connected impedances. (b) Wye-connected impedances.

tageous to replace the three delta-connected impedances by another set connected in wye (or star), Fig. 2-38b. If the new set has values properly selected, the behavior at the terminals *A*, *B*, and *C*, and the performance of the remainder of the system, will be unaffected by the substitution. It can be shown that the arrays of impedances are equivalent if

$$Z_A = \frac{Z_{ab}Z_{ca}}{Z_{ab} + Z_{bc} + Z_{ca}}$$

$$Z_B = \frac{Z_{bc}Z_{ab}}{Z_{ab} + Z_{bc} + Z_{ca}} \tag{2-110}$$

$$Z_C = \frac{Z_{ca}Z_{bc}}{Z_{ab} + Z_{bc} + Z_{ca}}$$

or

$$Z_{ab} = \frac{Z_A Z_B + Z_B Z_C + Z_C Z_A}{Z_C}$$

$$Z_{bc} = \frac{Z_A Z_B + Z_B Z_C + Z_C Z_A}{Z_A} \tag{2-111}$$

$$Z_{ca} = \frac{Z_A Z_B + Z_B Z_C + Z_C Z_A}{Z_B}$$

Sometimes the arrays of impedances discussed above are shown in a slightly different fashion, as indicated in Fig. 2-39. When presented in this manner the arrangements are sometimes designated as π-connected and *T*-connected, because of the diagram similarity to these two letters.

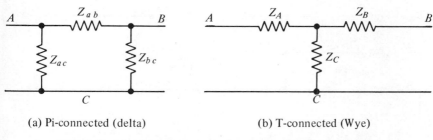

(a) Pi-connected (delta)   (b) T-connected (Wye)

FIG. 2-39. Circuits that are equivalent in relationship to terminals
A, B, and *C.* (a) π-connected (delta). (b) T-connected (Wye).

2-7. THREE-PHASE CIRCUITS

a. Three-Phase System of Voltage. A three-phase sys-
tem of voltages is illustrated in Fig. 2-40. A set of coils, *ab*, *cd*, and *ef*, which
might be three windings in an alternator or three secondary windings of

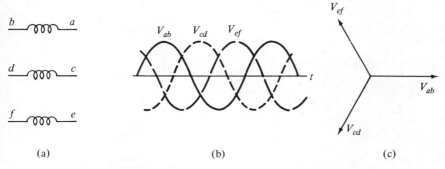

(a)   (b)   (c)

FIG. 2-40. A 3-phase system. (a) Coils in which voltages are
generated. (b) The sine wave voltages produced in the three coils.
(c) Phasor representation of the three voltages.

a transformer bank, is presented in (a). These coils are independent of
each other and may be interconnected as desired. The voltages generated
within these coils are presented as a time plot in (b) and as a phasor diagram
in (c). This system of voltages is said to be *balanced* when effective values
(or crest values) of the three waves are all of the same magnitude and the
time displacement between each is one third of a cycle (120 degrees).

b. Delta Connection. The coils of Fig. 2-40 may be connected
in delta as shown in Fig. 2-41 to form a three-phase supply with line conductors *X, Y,* and *Z*. In (a) the coils are shown as they would be connected physically. In (b) they are shown diagrammatically to emphasize the

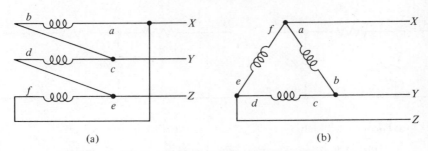

FIG. 2-41. Three coils connected in delta. (a) Physical arrangement.
(b) Diagrammatic representation.

delta connection. The voltages between each pair of line conductors X, Y, and Z are each equal in magnitude to the voltage of either of the coils *ab*, *cd*, or *ef*.

   c. Wye (or Star) Connection. The coils of Fig. 2-40a may be connected in wye (or star) as shown in Fig. 2-42 to form a three-phase supply with conductors X, Y, and Z. In (a) the coils are shown in physical

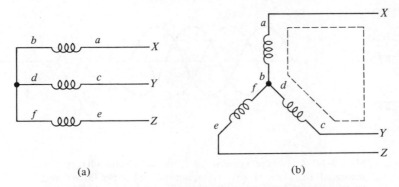

FIG. 2-42. Three coils connected in Wye. (a) Physical arrangement.
(b) Diagrammatic representation.

arrangement. In (b) they are shown diagrammatically to emphasize the wye connection. In a balanced three-phase wye system, the voltages between each pair of line conductors X, Y, and Z are equal in magnitude and are 1.73 times the voltage of either of the coils *ab*, *cd*, and *ef*. This voltage relationship may be demonstrated by applying Kirchhoff's voltage law around the path shown by the dashed line:

$$V_{ab} + V_{dc} + V_{yx} = 0$$

$$V_{ab} + V_{dc} = -V_{yx} = V_{xy}$$

The phasor sum of V_{ab} and V_{dc} (which is the same as the sum of V_{ab} and $-V_{cd}$) is shown in the phasor diagram, Fig. 2-43.

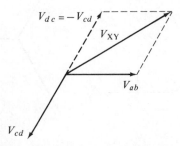

FIG. 2-43. Determination of the voltage between X and Y of the Wye connection, Fig. 2-42(b).

The common point of a wye connection is frequently termed the *neutral* and is designated by the letter N. The diagram of Fig. 2-42b might be shown as in Fig. 2-44.

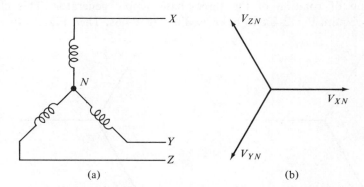

(a)   (b)

FIG. 2-44. The Wye-connected system. (a) The circuit showing the neutral. (b) The phasors of the line-to-neutral voltages. The phase sequence is XYZ.

d. Phase Sequence. The order in which each of the three voltages of a three-phase system reaches its positive crest determines the *phase sequence* of the system. Remembering that all phasors of a phasor diagram are assumed to rotate counterclockwise and that the positive crest of the sine wave is assumed to occur when the phasor is horizontal and to the right, it follows that the voltages of the system of Fig. 2-44b come up in the order X, Y, and Z. The phase sequence is therefore XYZ.

If the diagram of Fig. 2-42b had been labeled as shown in Fig. 2-45a (the designation X, Y, and Z is purely arbitrary), the phasor diagram would

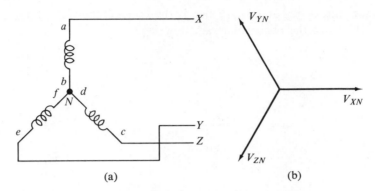

(a)   (b)

FIG. 2-45. The Wye-connected system with two terminals interchanged. (a) The circuit. (b) The phasors of the line-to-neutral
voltages. The phase sequence is ZYX.

appear as in Fig. 2-45b. Thus it is seen that interchanging two leads (such
as X and Y) reverses the phase sequence. Without interchanging connections, the phase sequence of a system may also be reversed by reversing the
direction of rotation of the three-phase supply generator. This change
would require Fig. 2-40c to be revised to Fig. 2-46a. Then Fig. 2-44b would

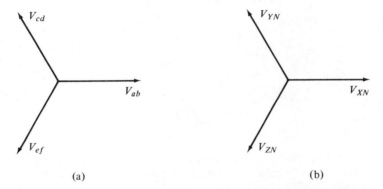

(a)   (b)

FIG. 2-46. Three-phase systems of phasors with the supply generator reversed in direction of rotation. (a) The phasors of coil
voltage. (b) The phasors of line-to-neutral voltages. The phase
sequence is ZYX.

be revised to Fig. 2-46b and the phase sequence would be ZYX. It must be
concluded that the phase sequence of a system can be determined only after
arbitrary conductor designations have been assigned.

   e. **Three-Phase Loads.** Three-phase loads are usually connected to the supply lines in delta or in wye as shown in Fig. 2-47. If the

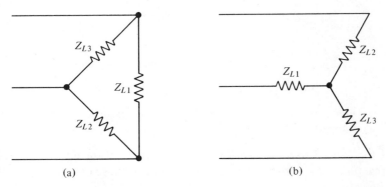

FIG. 2-47. Three-phase loads. (a) Delta connected. (b) Wye connected.

three impedances Z_{L1}, Z_{L2}, and Z_{L3} are identical, the loads are said to be *balanced.* When balanced three-phase loads are connected in delta, the line current is $\sqrt{3}$ times the current in an individual load impedance. In wye-connected systems, the line-to-line voltage is $\sqrt{3}$ times the voltage across any one individual load.

The power in a balanced three-phase load (wye or delta) is given by

$$P_{3\phi} = \sqrt{3}\ V_{LL}I_L \cos \phi \tag{2-112}$$

where V_{LL} is the line-to-line voltage, I_L is the line current, ϕ is the angle between the current and voltage in any one of the load impedances Z_L, and $\cos \phi$ is the power factor F_p of the load. The vector power of a balanced three-phase load is

$$P_{s3\phi} = \sqrt{3}\ V_{LL}I_L \qquad \text{volt-amperes} \tag{2-113}$$

2-8. EQUIVALENT CIRCUITS OF MACHINES

Many of the machines used in the electric power industry may be studied in regard to their performance in association with other equipment by replacing them with equivalent electric circuits. These electric-circuit models of the machines are usually developed on the basis of certain simplifying assumptions. These assumptions must be kept in mind when applying the models to the solution of power-system problems.

a. A C G e n e r a t o r . The circuit model of an ac generator (or alternator) is shown in Fig. 2-48. The voltage V_g is a fictitious generated voltage which is assumed to be dependent only on the speed of the machine and the value of the field current. The terminals PQ are the load terminals

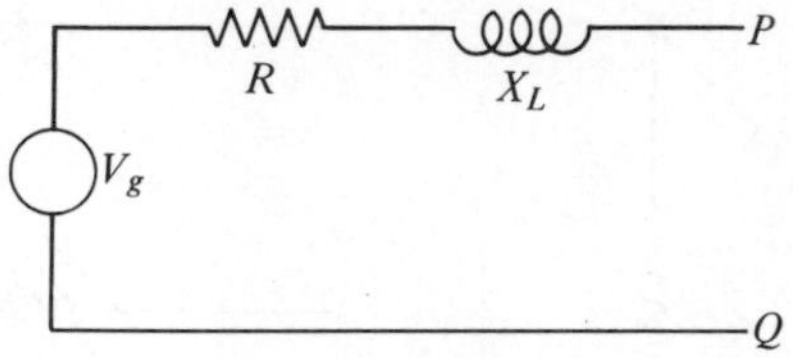

FIG. 2-48. The approximate equivalent circuit of an ac generator (or synchronous motor).

of the machine. The resistance R is the ac resistance of the machine winding. The reactance X is also fictitious, and its value is chosen to fit the problem, as follows:

X_d is termed the *synchronous reactance*, and this value is used in problems involving steady-state analysis.

X'_d is termed the *transient reactance*, and is used for problems involving sudden changes on the machine wherein numerical values are to be determined for a time approximately 10 cycles after the change is initiated.

X''_d is termed the *subtransient reactance*, and is used for problems involving sudden changes on the machine wherein numerical values are to be determined immediately after the change is initiated.

The different values of X are necessary because armature reaction does not follow immediately with the establishment of armature currents. Time is required for the generator field to change as a result of a change in the armature current.

b. Synchronous Motor. The circuit model of a synchronous motor is essentially the same as that of an ac generator. In Fig. 2-48, P and Q are the supply terminals and V_g is the internal voltage generated within the machine.

c. Two-Winding Transformer. The circuit model of a two-winding transformer is shown in Fig. 2-49a. Here R_1 and X_1 represent the primary winding resistance and leakage reactance, and R_2 and X_2 are the corresponding values for the secondary winding. The core-loss current and the magnetizing current are accounted for by the currents through R_{CL} and X_M. The coupling between primary and secondary is assumed to be through the medium of a perfect transformer of turns ratio N_2/N_1.

For many purposes, a simpler model, Fig. 2-49b, may be used. Here the core-loss and magnetizing currents are neglected and the secondary impedance has been transferred to the low-voltage side and combined with

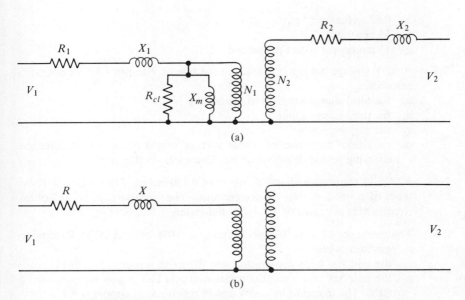

FIG. 2-49. The equivalent circuit of a two-winding transformer.
(a) The complete representation. (b) A simplified circuit.

the primary impedance. Obviously R and X could be transferred to the secondary side (if desired) by multiplying by $(N_2/N_1)^2$.

PROBLEMS

2-1.   Express the following quantities in the appropriate units of the MKS system.
    a.   15 inches
    b.   6 pounds mass
    c.   18 pounds force
    d.   60 miles per hour
    e.   3.5 days
    f.   16.2 square inches
    g.   32 pound feet torque
    h.   45 foot pounds of work
    i.   50 horse power
    j.   18,000 Maxwells
    k.   16 million electron volts

2-2.   Express the following in terms of English units.
    a.   4.2 kilometers
    b.   650 joules
    c.   85 kilowatts
    d.   55 meters per second

e. 0.42 webers per square meter

f. 65 newtons

g. 15 meters per scond per second

2-3.   A 12-V storage battery is charged at the rate of 9 amperes for 24 hours; calculate

a. the total charge supplied to the battery.

b. the total energy input.

c. the ampere-hours.

d. the size of the capacitor whose voltage would reach 12 volts after the charging period described above. Comment on this size.

2-4.   A No. 10 copper wire has a diameter of 0.100 inches. Determine the resistance of 6 miles of No. 10 copper wire. The resistivity, ρ, is 10.2 ohms circular mils per foot, or 1.7 10^{-8} ohm-meter.

2-5.   The resistance of a transformer winding is 0.076 ohm at 20°C. Determine its resistance when

a. the unit has been deenergized and stored in a room at −30°F.

b. the unit has been operated under overload and is at a temperature of 80°C. The temperature coefficient of resistance of copper is 3.9 $\times$ 10^{-3} per degree C.

2-6.   Three 6-ohm resistors are connected in series. Three identical 6-ohm resistors are connected in parallel. The series-connected group is then connected in series with the parallel-connected group. What is the resistance of the total assembly?

2-7.   Refer to problem 2-6. Suppose the parallel group of resistors is connected in parallel with the series group of resistors. Calculate

a. the resistance of the combination.

b. the conductance of the combination.

2-8.   A toroid-shaped iron core has a cross section 6 square centimeters and a mean diameter of 18 centimeters. The relative permeability is 970. A 120-turn coil wound on the core carries a current of 3.8 amperes. Calculate the magnetic flux in the core.

2-9.   The pole faces of a relay have an area of 1.5 square inches. By establishing a magnetic field between them it is desired to set up a force of 14 ounces. What must be

a. the magnetic flux density?

b. the total magnetic flux across the gap (neglect fringing)?

2-10.   A 350-turn coil is wound on an iron core. When the current through the coil is 5 amperes, the flux in the core is 0.6 webers. What is the self-inductance of the coil?

2-11.   Refer to Problem 2-10. A second coil of 200 turns is wound on the same core as described. What is the mutual inductance between the two windings?

2-12.   A capacitor has a plate area of 24 square feet and an insulation thickness of 0.035 inch. The relative dielectric constant is 4.2. The insulation may be operated at a stress of 350 volts per mil. Determine

    a. the capacitance of the capacitor.

    b. the voltage at which it may be operated.

    c. the energy stored in it when operating at full voltage.

2-13. The voltage between two conductors of a transmission line is 765 kilovolt rms. What is the crest (maximum) value of the voltage? Calculate the value of this voltage

    a. averaged over a positive half-cycle.

    b. averaged over a complete cycle.

2-14. Two sine-wave voltages are added together. One voltage has an rms value of 75,000 volts, the other 50,000 volts. What will be the value of their sum if the first voltage leads the second by

    a. $0°$?

    b. $60°$?

    c. $90°$?

    d. $180°$?

    e. $270°$?

2-15. A resistor of 6 ohms has applied across it a voltage of 500 volts. Calculate the current through the resistor if the frequency is

    a. 60 hertz.

    b. 60,000 hertz.

2-16. An inductor of 0.6 henry has 500 volts applied across it. Determine the current if the frequency is

    a. 60 hertz.

    b. 60,000 hertz.

2-17. A capacitor of 6×10^{-6} farad has 500 volts applied across it. Determine the current if the frequency is

    a. 60 hertz.

    b. 60,000 hertz.

2-18. A 6-ohm resistor and an 8-ohm inductive reactor carry a 60-hertz current of 550 amperes. What is the total voltage across the series combination? Draw a phasor diagram showing the current, the voltage across each part, and the total voltage.

2-19. A 60-ohm resistor, a 50-ohm capacitive reactor, and a 30-ohm inductive reactor are connected in series. The series combination is connected to a 480-volt 60-hertz supply. Calculate

    a. the total series impedance.

    b. the value of the current.

    c. the voltage across each unit.

    d. the power dissipated in each unit.

    e. the total power.

    f. the power factor.

    g. Draw a phasor diagram showing the current, the voltage across each unit, and the total voltage.

2-20. A resistor of 12 ohms and an inductive reactor of 16 ohms are connected across a 480-volt supply. Determine the power, the apparent power, and the reactive power supplied to the series combination.

2-21.   Given the circuit shown, using Kirchhoff's voltage law, write the equations for the voltages around each mesh. From these equations, solve for I_1 and I_2.

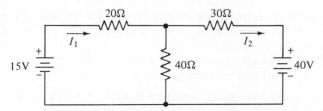

2-22.   Refer to Problem 2-21. Determine the current I_1 and I_2 using the method of superposition.

2-23.   For the circuit shown, determine the Thévenin equivalent circuit applying to the terminals *mn*.

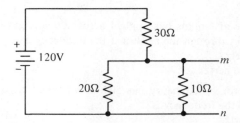

2-24.   Explain how you could determine experimentally the Thévenin equivalent of the convenience outlet in your study room.

2-25.   A group of three resistors are supplied by a balanced three-phase source as shown. Determine
   a.   the current in each resistor.
   b.   the current in each line conductor.
   c.   the total power supplied to the load.

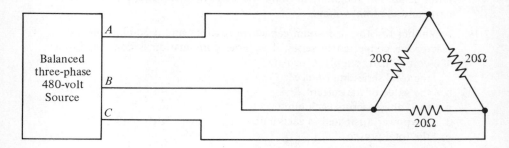

2-26. Solve for the current in each branch of the load (magnitude and phase angle). Determine the current in line AA'.

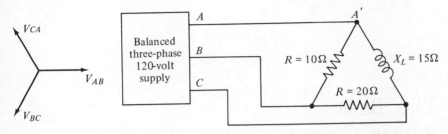

2-27. Repeat Problem 2-24 but assume that the phase sequence of the source is reversed, as shown in the accompanying phasor diagram.

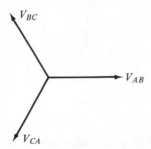

2-28. Solve for the current in each branch of the load. Determine the current in line NN'.

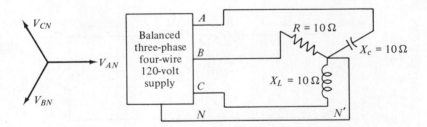

2-29. Repeat Problem 2-24 assuming the phase sequence CBA.

2-30. A large generator is rated as follows: 300,000 kilovolt-amperes, 22 kilovolt, three phase, 0.8 power factor. Determine
   a. the rated full-load current.
   b. the maximum power output of the machine.

chapter 3

The Basic
Power Circuit

3-1. THE BASIC CIRCUIT AND SOME APPLICATIONS

A simple circuit that has many applications in analyzing the behavior of electric power systems is shown in Fig. 3-1. This is a simple, single-phase

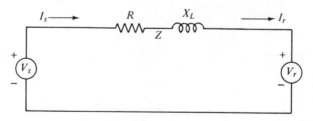

FIG. 3-1. The basic power circuit.

ac circuit consisting of a source voltage, V_s; an impedance, Z; and a receiver voltage, V_r. This simple circuit is worthy of note because it is the equivalent circuit representation of a number of different devices and serves as the basis for the analysis of a number of different problems associated with electric power supply.

This circuit may be used in studying problems of voltage regulation in which the supply voltage, V_s, is held constant, the load current is varied, and the problem is to study the variations in the receiver voltage, V_r, that

64

will result. A somewhat similar problem would be to specify that the receiver voltage V_r is to be held constant while the load current varies, with the sending voltage, V_s, specified as the unknown. In both problems it may be demonstrated that the power factor of the load will be of great significance. Losses in the power circuit and hence transmission efficiency may be calculated by reference to the resistance of the circuit.

With the behavior of this simple system clearly in mind, it is possible to develop important relations pertaining to the maximum power that may be transmitted over the circuit, the mechanical displacement between rotors of synchronous machines at the two ends of the circuit, and the magnitude of fault currents which may result when short circuits develop on the system.

3-2. SYSTEMS TO WHICH THE BASIC CIRCUIT APPLIES

The diagram of Fig. 3-1 presents the circuit model of a number of important systems in the electric power field. It will be seen that the devices to which this circuit may be applied are quite diverse in nature. The values of R and X that must be used in a specific problem will be discussed in Chapter 4.

a. Extension Cord. Although an ordinary extension cord is seldom regarded as a device on which electric-circuit calculations need to be made, its equivalent circuit will be found to be that shown in Fig. 3-1. The voltage V_s is the source voltage at the terminals of the convenience outlet into which the cord is attached. The voltage V_r is the voltage at the load end of the extension cord. The resistance R is the resistance of the two conductors making up the extension cord, while the inductive reactance X_L is that which results from the magnetic field linking between the two conductors of the cord when current flows to the load.

b. Power Lines. The behavior of transmission lines and distribution circuits is frequently analyzed by the short-line method represented in the circuit of Fig. 3-1. Although most transmission lines and distribution lines are three-phase circuits, they may be analyzed on a single-phase basis, which justifies the use of this simple circuit. Here V_s is the voltage at the sending end of the line and V_r is the voltage at the receiving, or load, end of the line. The justification for using a single-phase model for solving a three-phase circuit problem will be discussed in more detail later in this chapter.

c. AC Generator. As was described in Chapter 2, the equivalent circuit of an ac generator is that shown in Fig. 2-48. This is seen to be the same circuit as that shown in Fig. 3-1, in which V_s has replaced V_g and V_r has replaced the voltage at the terminals PQ.

d. Synchronous Motor. The equivalent circuit of the synchronous motor is a modification of the basic circuit Fig. 3-1, in which V_s is now the voltage at the supply terminals and V_r has become V_g, the voltage generated within the machine.

e. Transformer. The circuit model of the transformer was discussed briefly in Chapter 2 and presented in Fig. 2-49a. If the core loss and magnetizing current branches R_{CL} and X_M are neglected, the circuit of Fig. 2-49b results. This circuit is similar to the basic circuit (Fig. 3-1) except for the perfect transformer at the output terminals. If Fig. 3-1 is used for a study of transformer behavior, the secondary voltage becomes $V_r(N_2/N_1)$ and the secondary current becomes $I(N_1/N_2)$.

When a transformer is represented by *per unit quantities* (to be discussed in Chapter 4), the circuit representation of Fig. 3-1 may be used directly.

f. Series Combinations of Several Devices. In actual power systems it is very common to find that several of the components previously discussed are connected in series or in parallel combination. A series arrangement of an ac generator, an overhead power circuit, and a synchronous motor are shown in Fig. 3-2. As the impedances representing

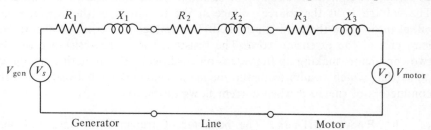

FIG. 3-2. Three basic power circuits in series.

them are in simple series combination, they may be added to form a single impedance extending from V_s, the internal generated voltage of the alternator, to V_r, the internal generated voltage of the motor.

g. Series–Parallel Combinations. The parallel combination of several devices, each represented by a simple impedance, may in turn be represented by a single equivalent impedance. Hence parallel combinations or series-parallel combinations of the circuit models of devices just discussed may be reduced to a simple circuit, as represented by Fig. 3-1.

h. Thévenin Equivalent of a Circuit. A complicated network such is that shown in Fig. 2-37a may have a load connected across two of its terminals such as *m* and *n*. For a study of the behavior of the load, the Thévenin equivalent of this system (Fig. 2-37d) may be used. This is another version of the basic circuit (Fig. 3-1).

3-3. EQUATIONS OF THE BASIC POWER CIRCUIT

The performance of the basic power circuit will be analyzed by phasor equations, by phasor diagrams, and by transmission diagrams. An interpretation of these diagrams will be further advanced in a later chapter. It may be noted from Fig. 3-1 that I_r and I_s are identical. For this reason there is no point in making a distinction between the two, and they will therefore be referred to simply as I.

Writing Kirchhoff's voltage equation around the basic circuit (Fig. 3-1) there results

$$\bar{V}_s - IZ - \bar{V}_r = 0$$

This equation may be solved for V_s:

$$\bar{V}_s = \bar{V}_r + IZ \tag{3-1}$$

It may also be solved for V_r:

$$\bar{V}_r = \bar{V}_s - IZ \tag{3-2}$$

Equations (3-1) and (3-2) may be regarded as the performance equations of the basic power circuit. They will be further analyzed in subsequent discussions.

3-4. PHASOR DIAGRAMS

The analysis of the basic circuit may begin by defining load-end conditions as V_r and I, as represented in the phasor diagram Fig. 3-3a. Here it may be noted that V_r is used as reference and so is drawn horizontally and to the right. The current I is shown *lagging* the receiver voltage by the angle ϕ_r, implying that the load is a combination of resistance and inductive reactance. The power factor of the load is equal to the cosine of ϕ_r.

In Fig. 3-3b all terms of Eq. (3-1) are displayed. The IZ drop is shown leading the current I by the angle β, where β is the angle whose tangent is X/R, sometimes referred to as the angle of the line. The IZ drop is then

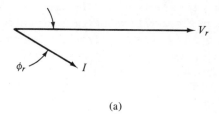

(a)

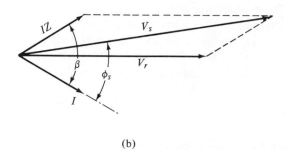

(b)

FIG. 3-3. Phasor diagram of the basic power circuit. (a) Receiver voltage and current. (b) The IZ drop added to receiver voltage gives sending-end voltage.

added by phasor addition to V_r to produce V_s. The angle ϕ_s is the angle between the sending-end voltage and the current. The power factor at the source terminals is given by cosine ϕ_s. Obviously the magnitude of the IZ drop is proportional to the magnitude of the current. It is also proportional to the magnitude of the line impedance.

The current may be divided into two components, one component I_p in phase with V_r and the other component I_q at right angles to V_r, as shown in Fig. 3-4a.

The line impedance is also made up of two components, the resistance R and the reactance X. The contribution to the IZ drop produced by the in-phase component of the current I_p flowing through the impedance is shown in Fig. 3-4b. The resistance drop is in phase with I, and the inductive reactance drop leads I_p by 90°.

The contribution to the IZ drop occasioned by the quadrature component of current I_q is shown in Fig. 3-4c. The IR drop is in phase with this component of current and the IX drop leads it by 90°.

The influence of the four components of the IZ drop taken collectively is shown in Fig. 3-4d. Here it may be noted that the two components I_pR and I_qX add directly in line with reference and so contribute substantially

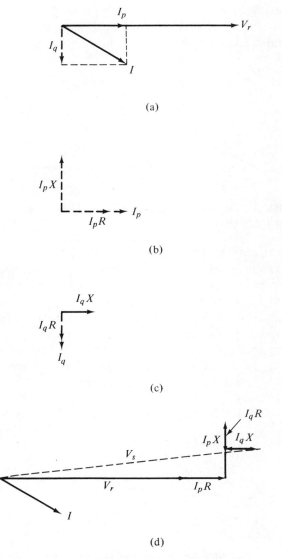

(a)

(b)

(c)

(d)

FIG. 3-4. Detailed phasor diagram of Eq. (3-1). Receiver current lagging receiver voltage. (a) Current divided into components I_p and I_Q. (b) The $I_p R$ and $I_p X$ drops. (c) The $I_q R$ and $I_q X$ drops. (d) The total diagram showing V_s.

toward an increase of the magnitude of V_s. The component $I_p X$ adds at right angles to V_r and the component $I_a R$ subtracts at 90° from V_r. These two components therefore tend to cancel each other. Since their resultant is at right angles to V_r, it adds but little in increasing the magnitude of V_s.

Figure 3-5 represents a condition similar to that just described except that here the current to the load *leads* the receiver voltage by the angle ϕ_r. This situation would apply if the load consisted of resistance and capacitive

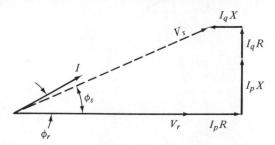

FIG. 3-5. Detailed phasor diagram of Eq. (3-1). Receiver current leading receiver voltage.

reactance. The diagram shows the four components of the IZ drop drawn in proper relation to the receiver voltage. It may be noted that I_pX and I_qR add at right angles to V_r and so make a limited contribution to the magnitude of V_s. The term I_pR adds directly to V_r, in contrast the term I_qX subtracts from it. Under these circumstances it is quite possible to have the magnitude of V_s smaller than the magnitude of V_r.

Turning to Eq. (3-2) it may be seen that the receiver voltage may be considered as the difference between the sending-end voltage and the IZ drop. When drawing a phasor diagram representing this equation, it is convenient to use V_s as reference and then to subtract the IZ drop from it, as shown in Fig. 3-6. In this particular diagram I is shown *lagging* V_s. By

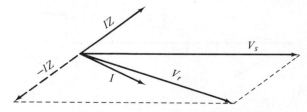

FIG. 3-6. Phasor diagram of Eq. (3-2).

appropriate changes, a similar diagram could be drawn in which I leads V_s.

Figure 3-7 shows a diagram similar to that of Fig. 3-6 except that the four components contributing to the IZ drop are clearly displayed. For a lagging power factor, V_r is always smaller in magnitude than V_s.

Figure 3-8 is a phasor diagram of Eq. (3-2) for the condition in which the current *leads* the sending-end voltage. The contributions of the four components of the IZ drop are shown on the diagram. Note that with a

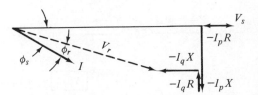

FIG. 3-7. Detailed phasor diagram of Eq. (3-2). The current I lags sending-end voltage V_s.

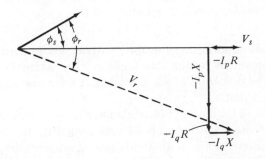

FIG. 3-8. Detailed phasor diagram of Eq. (3-2). The current I leads sending-end voltage V_s.

leading load and an inductive reactance in the line, it is possible for the receiver voltage to be greater in magnitude than the sending-end voltage.

Figures 3-3 to 3-8 were all drawn assuming that the impedance of the basic circuit (Fig. 3-1) was a resistance in series with an inductive reactance. Similar diagrams could be drawn for the situation in which the impedance is a resistance in series with a capacitive reactance. Capacitive impedances are seldom found in electric power circuits of the types discussed in Section 3-2.

3-5. TRANSMISSION DIAGRAMS

The performance of the basic power circuit for *any* condition of receiver current loading may be represented by an extension of the phasor diagrams just discussed. A chart known as a *transmission diagram* permits the graphical determination of many of the responses of the basic circuit. The transmission diagram is described by reference to a basic circuit to which particular numerical values have been assigned.

Consider the circuit shown in Fig. 3-9, in which the impedance, Z, is $5 \underline{/60°}\ \Omega$ and the receiver voltage V_r is 100 volts. Referring to Fig. 3-10, let V_r, drawn to scale, be the reference phasor. Assume that the current I is in phase with V_r as shown, indicating that the load power factor is

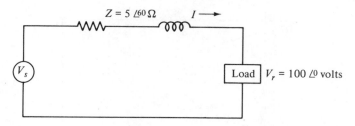

FIG. 3-9. The basic power circuit with the value of impedance Z and of receiver voltage V_r specified.

unity. On the assumption that the load current is 4 amperes, the IZ drop is properly represented by the line segment OA, which is 20 units long and lies along the line OM, which is located at a leading angle of 60° with respect to the current.

If the value of the current is now assumed to be 8 amperes and in phase with V_r, the IZ drop will be properly represented by the phasor OB (40 units long), as shown in Fig. 3-10. Similarly a 12 ampere current would give rise to an IZ drop 60 units in length, as represented by the phasor OC.

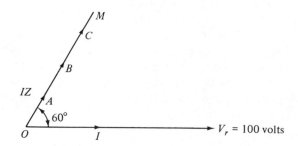

FIG. 3-10. Phasor diagram applying to Fig. 3-9 showing $V_r = 100 \, \underline{/0°}$ and the IZ drops for $I = 4 \, \underline{/0°}$, $I = 8 \, \underline{/0°}$, and $I = 12 \, \underline{/0°}$. (Scale: 1 cm = 20 volts.)

According to Eq. (3-1), V_s is determined by the phasor addition of V_r and the IZ drop. Consequently, the line OM may be shifted to the right until it adds vectorially to the phasor V_r, as shown in Fig. 3-11. For convenience we can label the points A, B, and C, in terms of the assumed currents 4, 8, and 12, as shown in Fig. 3-11.

Now a line drawn from O to A in Fig. 3-11 properly represents in magnitude and phase position the voltage V_s under a current loading of $I = 4 \, \underline{/0°}$. Similarly, a phasor drawn from O to the point marked 12 represents the correct value of V_s for unity-power-factor loading of 12 amperes.

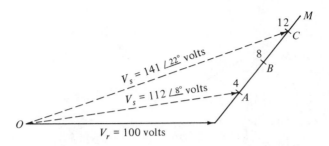

FIG. 3-11. The phasor diagram of Fig. 3-10 with the IZ drops added to V_r to give V_s.

We have constructed our transmission diagram for this system, although we are restricted to unity power factor at the load end of the line.

Let us now consider a zero-power-factor inductive load to be placed across the receiver terminal. The current I will now lag 90° behind the receiver voltage, V_r (Fig. 3-12). Again the IZ drop will be 60° ahead of the

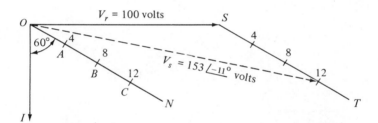

FIG. 3-12. Phasor diagram applying to Fig. 3-9 showing $V_r = 100 \, / \, \underline{0}$ and the IZ drops for $I = 4 \, / \underline{-90°}$, $I = 8 \, / \underline{-90°}$, and $I = 12 \, / \underline{-90°}$.

current and so will lie along the line ON as shown. Line ON may be scaled off to give points A, B, and C, again corresponding to load currents of 4, 8, and 12 amperes. As in the previous example, line ON may be shifted to the right so that it takes up the position ST from the end of the phasor V_r. Now a line drawn from O to the point 12 gives us the correct phasor value of V_s for the loading corresponding to a current of 12 amperes at zero power factor lag.

A similar diagram is shown in Fig. 3-13, except here the load current is assumed to be at 90° ahead of receiver voltage, V_r, indicating a pure capacitive load at the receiver end. The IZ drop scaled as before is drawn 60° ahead of the current phasor.

As before, the line OP may be transferred to the right so that it forms a phasor addition with V_r as shown by the line KL. Now a line drawn from

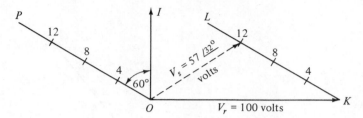

FIG. 3-13. Phasor diagram applying to Fig. 3-9 showing $V_r = 100\ /0°$ and the IZ drops for $I = 4\ /90°$, $I = 8\ /90°$, and $I = 12\ /90°$.

O to the point marked 12 properly represents the phasor value of V_s with a 12 amperes, zero-power-factor leading load at the receiver end.

Obviously Figs. 3-11, 3-12, and 3-13 may be combined in a single diagram, as shown in Fig. 3-14. Now we can determine the phasor value of V_s for *any* condition of receiver loading corresponding to power factor 1.0, power factor 0 lag, and power factor 0 lead.

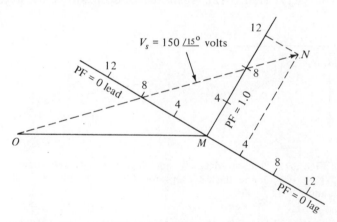

FIG. 3-14. Phasor diagrams applying to Fig. 3-9 for $V_r = 100\ /0°$ and the IZ drops for current of any value.

Next let us suppose that the receiver load consists of two parts, one that draws a current of 12 amperes in phase with receiver voltage and one that draws a current of 4 amperes, zero-power-factor lag. Each of these currents causes its own individual IZ drop, each of which can be identified on the diagram of Fig. 3-14. The distance from M to N properly represents this total IZ drop. We can now conclude that the phasor ON properly represents the sending-end voltage when the load current is $12 - j4$.

The discussion of the preceding paragraph suggests that we might further revise our diagram by drawing a *grid* in the IZ area as shown in Fig. 3-15. On this diagram the line ON properly represents the sending-end voltage V_s when the receiver current is $8 + j8$.

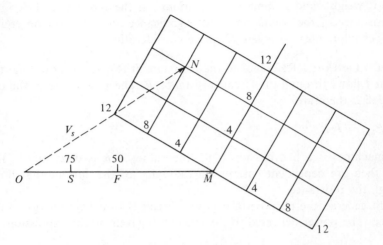

FIG. 3-15. The complete transmission diagram applying to Fig. 3-9 for $V_r = 100 \ \underline{/\,0°}$, $V_r = 75 \ \underline{/\,0°}$, and $V_r = 50 \ \underline{/\,0°}$.

Referring to Fig. 3-15, it may be noted that the area which has been divided into a grid is determined in grid *spacing* and in *angle* with respect to reference in accordance with the magnitude and angle of the impedance of the line. The length OM is determined by the magnitude of the receiver voltage V_r. Obviously this length would be only half as long if the receiver voltage had been 50 volts instead of 100 volts. We can therefore designate a point halfway between O and M (labeled on the diagram as F) and use this as the origin for problems in which the receiver voltage is assumed to be 50 volts. Another point, S, could be used as the origin of the diagram for problems in which the receiver voltage is assumed to be 75 volts. Other points along the line MO and its extension could be appropriately marked to be used as the origin of the diagram for problems involving any specified receiver voltage. Many other modifications of this diagram may be made.

It is left to the reader to work out the transmission diagram corresponding to Eq. (3-2). The usefulness of these diagrams will be apparent to the reader who works the problems given at the end of the chapter.

3-6. LOSSES AND THERMAL LIMITS

In the basic power circuit (Fig. 3-1) it may be noted that one of the components of the impedance is the circuit resistance. Resistance appears in all electric circuits (except those operated at extremely low temperatures) and is subject to limited control by the designer. The resistance of an electric conductor may be minimized by the use of low-resistivity materials, such as copper and aluminum, and by the selection of conductors of large cross-sectional area. The cross sections which may be used are frequently dictated by cost, weight, and volume considerations. In the most careful designs, the resistance of the conductors is of significance and its effect on equipment behavior must be given adequate consideration.

a. Losses. Reference to the basic circuit (Fig. 3-1) shows that the current I flows through the circuit resistance R. The power loss in the circuit itself is therefore

$$P_{\text{lost}} = I^2 R \qquad\qquad (2\text{-}8)$$

In a machine or a line already built, the value of R is determined. The losses then are dependent entirely on the value of the current that flows through the resistance.

In general, the purpose of a power circuit is to deliver energy to the receiver. The power delivered to the receiver is given by the expression

$$P_r = V_r I \cos \phi_r \qquad\qquad (3\text{-}3)$$

where ϕ_r is the angle between V_r and I. From this expression the current is seen to be

$$I = \frac{P_r}{V_r \cos \phi_r} \qquad\qquad (3\text{-}4)$$

and

$$P_{\text{lost}} = \left(\frac{P_r}{V_r \cos \phi_r}\right)^2 R \qquad\qquad (3\text{-}5)$$

Usually the voltage V_r is dictated by the rating of the load equipment and so must be held substantially constant. In some instances there is opportunity for the control of the power factor of the load. Obviously, the current I has its smallest possible value when the power factor is unity, a situation which exists when receiver voltage and current are in phase and the power factor angle ϕ_r is zero.

Many electrical machines (such as induction motors) contain iron which must be magnetized for the machines to function. The magnetizing currents, which lag the receiver voltage by 90°, are almost constant regardless of motor loading. The in-phase components of currents taken by motors increase as the motors are loaded. Hence the power factor of induction motor loads is less than unity and always lagging. It is possible to compensate for the quadrature lagging current by installing capacitors in parallel with the motor load. If the quadrature leading current of the capacitor is exactly equal to the quadrature lagging current of the motors, the power factor can be made equal to unity. At this unity power factor, the magnitude of the line current I is a minimum and losses are at a minimum.

Synchronous motors provide a means of regulating load power factor. By control of the magnitude of the field current supplied to the synchronous motor, the machine may draw a current having either a leading or a lagging component. If the synchronous motor is the only load on the end of a power circuit, the adjustment of the field current to produce unity power factor will minimize the current and consequently the losses in the power circuit. If the synchronous motor is in parallel with induction motors, which are drawing a lagging power factor current, the synchronous motor may be operated with excess field current thus drawing a leading current from the line, which, in combination with the other lagging motor currents, may result in unity power factor.

b. Thermal Effects. The I^2R losses represent electrical energy converted into heat within the conductor whose resistance is being considered. The liberated heat tends to increase the temperature of the conductor; the temperature increase depending on the rate at which heat is liberated and the ability of the surrounding environment to transfer this heat away. In some cases, as in generators or transformers, forced circulation of gas or oil over the conductors is necessary to hold temperatures to reasonable values.

In practically all electrical equipment, consideration must be given to the maximum temperature at which the conductors operate. In electrical machines and in power cables, the conductors are separated from each other and from grounded objects by means of solid insulation. The dielectric materials which form the insulation are, in most instances, organic materials that undergo chemical changes when subjected to high temperatures. Under extremely high temperatures, these materials will quickly char and lose their ability to serve as insulators. Insulation failure results in short circuits, and the destruction of the equipment may follow.

Overhead lines, for the most part, are constructed of *bare* conductors suspended on porcelain insulators. On such lines, the temperature of the conductors has no detrimental effect on the insulators. However, most power-line conductors are made of alloys of steel, aluminum, or copper, which are carefully heat-treated to develop maximum mechanical strength. Operation of these conductors at high temperatures may anneal the metal and seriously reduce conductor mechanical strength.

Damage to the solid insulation of machines, such as generators and transformers, and to metals of overhead conductors depends on the temperature reached during operation. The *ambient temperature* (the temperature of the surrounding environment) is of great importance in the operation and rating of electrical equipment. For example, a motor operating in a location where the temperature is $-20°F$ may carry a much higher load without damage to its insulation than could a similar motor operating in

(3-9)

(3-10)

(3-11)

impedances Z_Δ
equivalent wye-
Z_Y (as show

(3-12)

ed above. Equa-
and delta-connected

KW, power factor 0.8
having a resistance of
factor (or line to neut

$$\frac{3,000,000}{\sqrt{3} \times 22,000 \times 0.8} = 98.6 \text{ A}$$

$/0°$ (reference)
$6(0.8 - j.6) = 79 - j59.2$

$= 829 + j613$

placed
the load
3-16.

a room where the temperature is 120°F. Electrical insul... tained below a certain temperature limit, regardles... to it.

Because electrical machines operate in l... ambient temperatures, different types of... temperature of the surroundings plu... operation must be given conside... electrical machine.

3-7. BALANCED T...

In the disc... made th... circu... F...

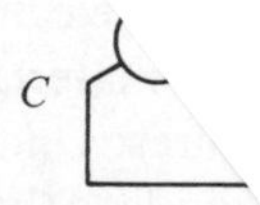

FIG. 3-16. L... through a 3-ph...

phase load, *abc*, by means... conductor of this line there is... The balanced three-phase load... Z_L, which are here shown as st... shown connecting the neutral of t... Since this is a balanced source supp... voltages and the three load currents... phasor diagram, as shown in Fig. 3-17. W... about the node *n* (Fig. 3-16) we find

$$\bar{I}_{Nn} + \bar{I}_{an} + \bar{I}_{bn} + \bar{I}_{cn} = 0$$

This method of handling a problem involving a balanced three-phase system is sometimes termed the *line-to-neutral method* of solution. The impedance used is the line-to-neutral resistance and reactance (or the resistance and reactance per conductor). In using this method the following relations may be noted

$$V_{LL} = \sqrt{3}V_{LN} \quad (3\text{-}8$$

$$P = 3V_{LN}I_L \cos\phi = \sqrt{3}V_{LL}I_L\cos\phi \quad (3\text{-}$$

$$I_L = \frac{P}{\sqrt{3}V_{LL}\cos\phi}$$

$$P_{\text{lost}} = 3I_L^2R$$

in which all terms are scalars.

Suppose that the load is made up of three identical imp... connected in delta. The delta load may be replaced by an equ... connected load made up of three impedances each of value... in Section 2-6e), where

$$Z_Y = \tfrac{1}{3}Z_\Delta$$

The new wye-connected load may be handled as descri... tions (3-8), (3-9), (3-10), and (3-11) apply both for wye-... loads.

EXAMPLE 3-1. A balanced three-phase load of 3000... is supplied at 22,000 V (line to line) over a line h... and an inductive reactance of 10 Ω per cond... Calculate the

a. line current.
b. phase angle.
c. line-to-neutral voltage.
d. impedance volt drop in line.
e. line-to-neutral source voltage.
f. line-to-line source voltage.
g. power lost in all three lines.

Solution. a. $I = \frac{P}{\sqrt{3}V_{LL}\cos\phi} = \ldots$

b. $\phi = \cos^{-1} 0.8 = 37°$

c. $\bar{V}_{LN} = \frac{V_{LL}}{\sqrt{3}} = \frac{22{,}000}{\sqrt{3}} = 12{,}7\ldots$

d. $\bar{I} = I(\cos\phi - j\sin\phi) = 98\ldots$

$Z = 3 + j10$

$\bar{I}Z = (79 - j59.2)(3 + j10\ldots$

e. $V_{S(\text{line to neutral})}$ $= \bar{V}_r + \bar{I}Z$
$= 12,700 + 829 + j613$
$= 13,530 + j613 = 13,530 \underline{/2.6°}$ (3-7)

f. $V_{S(\text{line to line})}$ $= \sqrt{3}\ V_{S(L-N)} = \sqrt{3} \times 13,530$
$= 23,400$ V line to line

g.   The total power lost is
$P = 3I_L^2 R$

$= 3(98.6)^2 \times 3 = 87,500$ W   (3-11)

3-8.   SINGLE-PHASE EQUIVALENT OF A THREE-PHASE SYSTEM

In Section 3-7 it was shown that a three-phase system made up of a balanced three-phase generator, a balanced line, and a balanced three-phase load could be studied by considering only one third of the system and so analyzing it on a line-to-neutral basis. The behavior of the three-phase system may also be analyzed using the single-phase equivalent circuit shown in Fig. 3-19. This circuit differs from Fig. 3-18 in that the voltages used at both

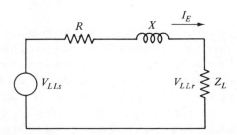

FIG. 3-19. The single-phase equivalent of the 3-phase system, Fig. 3-17.

the sending and receiving ends are line-to-line voltages instead of line-to-neutral voltages, as previously used. The impedance of the load, Z_L, at the receiver end is identical to that previously used, but because of the higher voltage impressed across it, the power consumed in this load is equal to the total three-phase load assumed in Fig. 3-16. The line impedance is identical to that in one of the three conductors.

In the single-phase equivalent circuit the line current is seen to be

$$I_E = \frac{P_{3\phi}}{V_{LLr} \cos \phi_r}$$ (3-13)

In comparison to Eq. (3-10) it may be seen that the current in the single-phase equivalent circuit, I_E is

$$I_E = \sqrt{3}\ I_L$$ (3-14)

The power lost in the equivalent circuit is

$$P_{\text{lost}} = I_E^2 R = 3I_L^2 R \qquad (3\text{-}15)$$

which is the same as in the actual circuit as given by Eq. (3-11).

From Fig. 3-19 we may write the relation

$$\bar{V}_{LLs} = \bar{V}_{LLr} + I_E Z \qquad (3\text{-}16)$$

If we multiply each term of Eq. (3-7) by the factor $\sqrt{3}$ we obtain

$$\sqrt{3}\,\bar{V}_{AN} = \sqrt{3}\,\bar{V}_{an} + \sqrt{3}\,I_{Aa}Z \qquad (3\text{-}17)$$

It may be noted that (in terms of scalar values)

$$\sqrt{3}\,V_{AN} = V_{LLs}$$

$$\sqrt{3}\,V_{an} = V_{LLr}$$

$$\sqrt{3}\,I_{Aa} = I_E$$

Hence the single-phase equivalent-circuit solution gives the correct value of the losses and the correct scalar value of the sending-end line-to-line voltage.

In summary, it may be stated that a system consisting of a balanced three-phase generator connected to a balanced three-phase load (either star- or delta-connected) through a three-phase line may be analyzed by a single-phase equivalent circuit. The voltages in this single-phase equivalent circuit are equal in magnitude to the corresponding line-to-line voltages of the actual three-phase circuit. The losses in the equivalent single-phase circuit are identical to the losses in the three-phase circuit. The current in the equivalent single-phase circuit has a magnitude equal to $\sqrt{3}$ times that of the line current in the actual three-phase circuit.

EXAMPLE 3-2. Solve Example 3-1 by the single-phase equivalent method. *Solution.*

$$P = 3000 \text{ kW}$$

power factor $= 0.8$ lag

$$V_r = 22{,}000 \text{ V} \qquad \text{(reference)}$$

$$Z = 3 + j10 = 10.5 \,\underline{/73.3^\circ}\ \Omega$$

$$I_E = \frac{P_{3\phi}}{V_{LL}\cos\phi}$$

$$= \frac{3{,}000{,}000}{22{,}000 \times 0.8} \qquad (3\text{-}13)$$

$$\check{I}_E = 171 \ \underline{/-37°} \ \text{A}$$

$$
\begin{aligned}
\check{V}_{LLS} &= \check{V}_{LLr} + \check{I}_E Z \\
&= 22{,}000 \ \underline{/0°} + (171 \ \underline{/-37°})(10.5 \ \underline{/73.3°}) \\
&= 22{,}000 \ \underline{/0°} + 1790 \ \underline{/36.3°} \\
&= 22{,}000 + 1440 + j1060 \\
&= 23{,}440 + j1060 = 23{,}450 \ \underline{/2.6°} \ \text{V}
\end{aligned}
\tag{3-16}
$$

The power lost is

$$
\begin{aligned}
P &= I_E^2 R \\
&= (171)^2 \times 3 = 87{,}500 \ \text{W}
\end{aligned}
\tag{3-15}
$$

Note that the calculated sending-end (line-to-line) voltage and the losses agree with the values found in Example 3-1.

3-9. TRANSMISSION-LINE EQUIVALENCE; A BETTER CIRCUIT MODEL

In Section 3-2 it was stated that the basic circuit (Fig. 3-1) was useful as the short-line representation of an electric power circuit. This representation is adequate for most overhead-distribution circuit problems and for many high-voltage transmission-line problems, particularly if the line length is less than 50 miles. It is inadequate, however, for some distribution circuit problems, for many high-voltage transmission-line problems (particularly for those involving circuits greater than 50 miles in length), and for problems involving cable circuits of even shorter length. In many cases where the short-line representation is inadequate, the π-line or T-line representation will give satisfactory results. For some advanced studies even these representations are inadequate.

The short-line method of solution is sometimes inadequate because it fails to represent the capacitance present in all power circuits. In some problems this capacitance is of little importance; in others it plays an important role. Two conducting bodies separated by a dielectric constitute a capacitor. An overhead transmission line consists of three power conductors supported above the ground, which itself is a conductor. In addition, there may be one or two ground wires. Obviously capacitance exists between every pair of these conducting bodies, giving a capacitance array that may be very complicated. In underground cables, the distance between conducting bodies is very short and the dielectric separating them has a permittivity several times as great as air. The capacitance per mile of underground cables is much greater than the capacitance per mile of overhead lines.

For the present discussion we shall consider only single-phase systems and balanced three-phase systems carrying balanced three-phase loads. For problems of this sort, the multicapacitor system associated with overhead lines and cables may be simplified to a circuit arrangement in which a single capacitor appears between each conductor and neutral. The capacitance, like the line resistance and inductance, is distributed throughout the line length, a portion appearing in every meter of the length. However, it may be shown by mathematical analysis (not included here) that the capacitance, the resistance, and the inductance may be *lumped* to form relatively simple equivalent circuits. One of these lumped-circuit arrangements is known as the π-line representation; the other is known as the T-line representation.

In the π-line representation, the circuit of Fig. 3-16 is modified to become the circuit shown in Fig. 3-20. It may be noted in this figure that

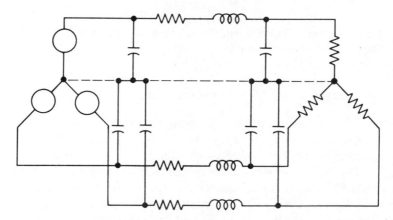

FIG. 3-20. A 3-phase system, showing the resistance, the inductance, and the capacitance of the transmission line. Pi-line representation.

a capacitor is shown between line and neutral on each of the three conductors at both ends of the circuit. The line-to-neutral representation or the single-phase equivalent of this circuit is shown in Fig. 3-21.

In the π-line representation (Fig. 3-21) the resistance and inductive reactance are represented exactly as in the short-line method. Here R represents the resistance and X_L the series inductive reactance of a single conductor for the length of the line. The line-to-neutral capacitance for the length of line is determined and half of this value is put at each end of the line. It should be noted in Fig. 3-21 that the sending-end current, I_s is not the same as the receiver-end current, I_r, and a distinction must be made between them.

The line-to-neutral (or the single-phase equivalent) of a transmission circuit represented by the T-line method is shown in Fig. 3-22. Here the total resistance and the total inductive reactance of the line have been divided into two parts. The total capacitance of the circuit is connected between

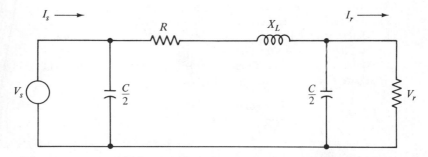

FIG. 3-21. The line-to-neutral (or single-phase equivalent) representation of the 3-phase system, Fig. 3-20. *R*, *L*, and *C* are the total circuit resistance, inductance, and capacitance per conductor. The Pi-line representation.

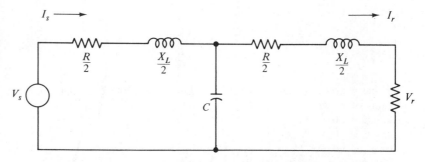

FIG. 3-22. The T-line representation of a transmission system.

line and neutral at the midpoint. Again I_s differs from I_r and must be accordingly designated.

The π and T representations of transmission lines will be discussed in more detail in later chapters.

In some advanced problems it is necessary to recognize that the line resistance, inductance, and capacitance is distributed throughout every meter of the entire line length. Relations may be worked out on this basis which provide an exact solution to transmission-line circuit behavior.

Problems

3-1.   Refer to Fig. 3-2. Assume that the generator has an impedance $Z_G = 0 + j20$, the line an impedance of $Z_L = 15 + j50$, and the motor an impedance $Z_M = 0 + j40$. Calculate the impedance of the basic circuit which may be considered for analysis.

3-2.   A power system is represented by an equivalent diagram as shown. All
       impedances have been referred to the same voltage level.
       a. Determine the basic circuit considering the generator bus voltage as V_s
           and the load bus voltage as V_r.
       b. Determine the basic circuit considering the generator internal voltage
           V_g as V_s and the load bus voltage as V_r.

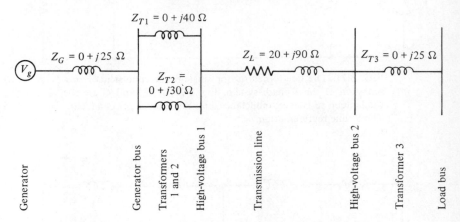

3-3.   A basic power circuit is as shown. Let $\bar{V}_r = 120\ \underline{/0°}$ V. Determine the
       value of V_s when
       a. $\bar{I}_r = 10\ \underline{/0°}$ A.
       b. $\bar{I}_r = 10\ \underline{/-90°}$ A.
       c. $\bar{I}_r = 10\ \underline{/90°}$ A.
       Draw a phasor diagram for each case.

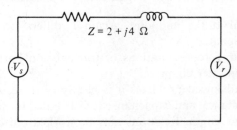

3-4.   Consider a basic power circuit in which $Z = 3 + j10\ \Omega$ and $\bar{V}_r = 240\underline{/0°}$ V.
       Determine the value of V_s if $\bar{I}_r = 4 - j8$ A.

3-5.   Consider the circuit in which $Z = 3 + j8\ \Omega$, $\bar{V}_r = 200\ \underline{/0°}$ V, and
       $P_r = 1200$ W. Determine the values of V_s and P_s when
       a. receiver power factor = 1.0.
       b. receiver power factor = 0.6 lag.
       c. receiver power factor = 0.6 lead.

3-6.   Consider the circuit in which $Z = 3 + j8\ \Omega$, $V_s = 240\ \underline{/0°}$ V, and
       $\bar{I}_s = 10\ \underline{/-30°}$ A. Solve for V_r and P_r. Draw a phasor diagram.

3-7.   Consider the circuit in which $Z = 3 + j8\,\Omega$, $\bar{V}_s = 240\,\underline{/0°}$ V, and $Z_r =$ (load impedance) $= 7 + j22\,\Omega$. Calculate
   a.   receiver current I_r.
   b.   receiver voltage V_r.
   c.   receiver power factor F_{pr}.
   d.   receiver power P_r.
   e.   receiver volt-amperes P_{sr}.
   f.   receiver reactive volt-amperes P_{qr}.

3-8.   Consider the circuit in which $Z = 3 + j8\,\Omega$, $\bar{V}_s = 240\,\underline{/0°}$ V, $P_r = 1000$ W, and $F_{pr} = 0.8$ lag. Solve for V_r (magnitude and angle).

3-9.   Consider the circuit in which $Z = 6 + j8 = 10\,\underline{/53°}\,\Omega$ and $\bar{V}_r = 500\,\underline{/0°}$ V. On a single phasor diagram solve graphically for V_s (magnitude and angle), for each of the following conditions:
   a. $\bar{I}_r = 20\,\underline{/0°}$ A.
   b. $\bar{I}_r = 20\,\underline{/90°}$ A.
   c. $\bar{I}_r = 20\,\underline{/-90°}$ A.
   d. $\bar{I}_r = 20\,\underline{/-45°}$ A.

3-10.   Draw a transmission diagram applying to a basic power circuit in which $Z = 20.0\,\underline{/75°}\,\Omega$. Use V_r as reference. Draw a grid line for each 50 A, from 0 to 300 A. On the V_r line, mark appropriate points for $V_r = 5000$, 10,000, and 15,000 V. Use a scale of 1 cm $= 1000$ V. From the diagram determine V_s (magnitude and angle) for each of the following conditions:
   a. $V_r = 10,000$ V; $I_r = 150$ A; $F_{pr} = 1.0$.
   b. $V_r = 15,000$ V; $I_r = 200 + j100$ A.
   c. $V_r = 10,000$ V; $P_r = 2000$ kW $F_{pr} = 0.8$ lag.

3-11.   Refer to the circuit of Problem 3-10. What is the largest load of 0.707 power factor lag which may be carried on the circuit while maintaining V_r at 12,500 V while V_s is limited to 16,000 V? For this load condition determine also P_s and the losses in the circuit.

3-12.   For the circuit of Problem 3-10, draw a transmission diagram using V_s as reference. Solve for V_r when
   a. $V_s = 10,000$ V; $I_s = 200$ A. $F_{ps} = 1.0$.
   b. $V_s = 15,000$ V; $\bar{I}_s = 150 + j200$.
   c. $V_s = 15,000$ V; $\bar{I}_s = 250\,\underline{/-30°}$.
   d.   With the answer to part c, refer to the diagram of Problem 3-10. Does this give you $V_s = 15,000$ V, as expected?

3-13.   A transmission circuit has an impedance of $Z = 3 + j14\,\Omega$. Load end conditions are $V_r = 5000$ V and $P_r = 250$ kW. Determine transmission losses and transmission efficiency for each of the four conditions:
   a.   Receiver current lags receiver voltage by 30°.
   b.   Receiver current lags receiver voltage by 60°.
   c.   Receiver current leads receiver voltage by 30°.
   d.   Receiver current is in phase with receiver voltage.

3-14. A three-phase transmission line has an impedance $Z = 6 + j20 \, \Omega$ per conductor. Load-end conditions are $V_{LLr} = 13,800$ V, P_r (three phase) = 1200 kW, and power factor = 0.8 lag. Solve for sending-end voltage, current, power factor, power, and transmission loss using the line-to-neutral method.

3-15. Repeat Problem 3-14 using the single-phase equivalent method.

chapter 4

Percent
and Per Unit
Quantities

4-1. INTRODUCTION

The use of the MKS system dictates that problems related to electric circuits should be solved in terms of volts, amperes, volt-amperes, and ohms. Answers to problems pertaining to electric power systems are almost always required in terms of these quantities. However, in the process of computations, it is frequently more convenient to express these electrical quantities in terms of "percent" or "per unit" of some arbitrarily chosen base. The electrical characteristics of machines are usually specified by the designers and manufacturers in terms of percent or per unit.

4-2. THE PERCENT SYSTEM

The *percent system* will be illustrated by means of an example. Consider an impedance Z (Fig. 4-1) which is assumed to have a constant value of 40 ohms regardless of the voltage impressed across it. Let us assume that we shall use this 40 ohms as reference and so declare it to be 100 percent impedance, as shown. The value of other impedances may then be specified as a certain percentage of this reference value.

Next, suppose that because of the design of this impedance or because of some operating practice, we declare the voltage associated with this

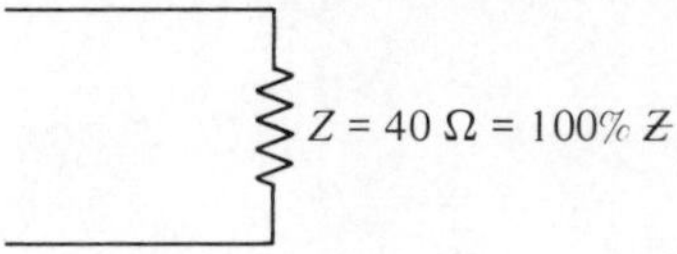

$$Z = 40 \ \Omega = 100\% \ Z$$

FIG. 4-1. A simple impedance, Z.

impedance to be arbitrarily set at 200 volts. That is, 200 volts will be declared to be 100 percent voltage. Since 200 volts (100 percent voltage) applied to the 40-ohm impedance (100 percent impedance) gives rise to a current of 5 amperes, it is logical to declare 5 amperes to be 100 percent current. Further, since 5 amperes at 200 volts implies 1000 volt-amperes, it follows that this value should represent 100 percent volt-amperes.

Those values which are declared to be 100 percent are termed *base values*. In summary, for this example,

base impedance	40 ohms	$100\% \ Z$
base voltage	200 volts	$100\% \ V$
base current	5 amperes	$100\% \ I$
base volt-amperes	1000 volt-amperes	$100\% \ VA$

Problems pertaining to conditions other than those described above might be specified in terms of these base quantities. A 50-ohm impedance would be specified as

$$Z = 125\%$$

a current of 4 amperes would be specified as

$$I = 80\%$$

and a voltage of 150 volts would be specified as

$$V = 75\%$$

The percent system presents some inconsistencies when values expressed in this fashion are inserted into mathematical equations. For example, we may write

$$V = IZ$$

If I is 80 percent (4 amperes) while Z is 150 percent (60 ohms), the relation above would yield

$$V = IZ = 80 \times 150 = 12,000\%$$

This result implies a voltage of 24,000 volts whereas the answer should be 240 volts.

Although it is common practice to express quantities in many instances as a percent, computations are much more readily made in a modification of the percent system, known as the *per unit system*.

4-3. THE PER UNIT SYSTEM

The *per unit system* is similar to the percent system except that all quantities are expressed as *decimal fractions* instead of percentages. Base quantities then have the value unity instead of 100 percent.

For the tabulation given in Section 4-2 we now write

base impedance	40 ohms	1.0 pu
base voltage	200 volts	1.0 pu
base current	5 amperes	1.0 pu
base volt-amperes	1000 volt-amperes	1.0 pu

It then follows that

50 ohms is 1.25 pu impedance
4 amperes is 0.8 pu current
150 volts is 0.75 pu voltage

The per unit system has the distinct advantage that, with it, all basic circuit relations apply. For example, if I is 0.8 pu (4 amperes) and Z is 1.50 pu (60 ohms), it follows that

$$V = IZ$$
$$= 0.8 \times 1.50 = 1.2 \text{ pu V}$$

It is now seen that 1.2 pu volts corresponds to 240 volts, the correct answer.

In summary, when expressed in per unit quantities,

$$V_{base} = I_{base} Z_{base} \tag{4-1}$$

$$VA_{base} = V_{base} I_{base} \tag{4-2}$$

If any two of the four quantities in these equations are specified, the other two may be evaluated immediately. In most instances, base voltage and base volt-amperes are specified. Base current and base impedance may then be calculated. Actual values and per unit values are related by

$$Z_{\text{ohms}} = Z_{\text{pu}} Z_{\text{base}} \qquad\qquad (4\text{-}3)$$

$$I_{\text{amps}} = I_{\text{pu}} I_{\text{base}} \qquad\qquad (4\text{-}4)$$

$$V_{\text{volts}} = V_{\text{pu}} V_{\text{base}} \qquad\qquad (4\text{-}5)$$

$$VA_{\text{volt amps}} = VA_{\text{pu}} VA_{\text{base}} \qquad\qquad (4\text{-}6)$$

The per unit system simplifies many of the problems of circuit analysis. In the conventional form of calculation using volts and amperes, the solution of a system involving power lines of several different voltage levels requires all impedances which are to be added to be transferred to a single voltage level. In the per unit system, the different voltage levels entirely disappear and a power network involving generators, transformers, and lines (of different voltage levels) reduces to a system of simple impedances, each one of which is essentially the general power circuit discussed in Chapter 3. Furthermore, machines such as generators and transformers, when described in the per unit systems, have their characteristics specified by almost the same number, regardless of the rating of the machines. For example, a 20-kVA transformer may have 6 percent (0.06 pu) reactance and a 200,000-kVA transformer may have 6 percent (0.06 pu) reactance.

4-4. SPECIFICATION OF GENERATORS

Consider the equivalent circuit of a generator as discussed in Section 2-8a and as represented in Fig. 4-2. Let it be. assumed that the rating of the generator is

      1000 VA   200 V

From this we may calculate the rated single-phase current as

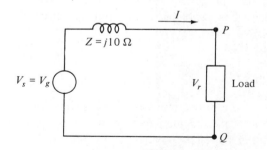

FIG. 4-2. The equivalent circuit of a generator.

$$I = \frac{VA}{V} = \frac{1000}{200} = 5 \text{ A}$$

The load impedance necessary to produce this rated current and rated volt-ampere output at rated voltage is

$$Z_L = \frac{V_r}{I} = \frac{200}{5} = 40 \ \Omega$$

For this machine we can define base quantities as follows:

base volt-amperes	1000 VA
base volts	200 V
base current	5 A
base impedance	40 Ω

For problems associated with this machine, volt-amperes, voltage, current, and impedance may be expressed as a percentage of base values or as a decimal fraction of base value (per unit).

From Fig. 4-2 it may be noted that the internal impedance of the generator is shown as 10 ohms. This is 25 percent, or 0.25, of Z_B. Hence we could say that on the basis of the machine rating, the internal impedance of the generator is 25 percent or 0.25 per unit.

Let us again examine the generator, but this time let us assume that the load terminals are short-circuited as shown in Fig. 4-3. Let us now ask

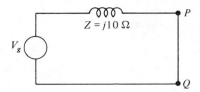

FIG. 4-3. The generator of Fig. 4-2 with its terminals short-circuited.

the question: What value of V_g is necessary to produce rated full-load current in the machine? Since the rated full-load current is 5 amperes and the impedance of the machine is 10 ohms, it follows that

$$V_g = IZ = 5 \times 10 = 50 \text{ V}$$

Thus it is seen that the voltage required is 25 percent of the *rated* terminal voltage. Expressing this as a decimal fraction, we may say that the voltage required to circulate full-load current is 0.25 times rated voltage. In the terminology of per unit quantities we say that this voltage is 0.25 per unit.

It may be noted that the voltage required to circulate full-load current with the machine short-circuited at the terminals expressed in per unit is exactly equal to the impedance of the machine expressed in per unit.

The rating of a machine, such as a generator, is determined not only by the design of the machine but also by the condition of operation to which it is to be subjected. Suppose, for example, that the machine of Fig. 4-2 is to be used in an application where it will never be run more than 5 minutes during any 1-hour period. For this short-time operation, the machine might easily be rated 2500 volt-amperes. However, since nothing else is changed, the equivalent circuit of Fig. 4-2 still applies, for the ohms impedance of the machine is unaffected by the nameplate specifications. Under this new condition of operation we can now write

base volt-amperes	2500 VA
base voltage	200 V
base current	12.5 A
base impedance	16 Ω

The per unit Z of the machine is now 10 divided by 16, the actual impedance divided by the base impedance, or 0.625 per unit. Any numerical problems involving volts, volt-amperes, amperes, and ohms (not per unit quantities) worked out by the new base quantities should give the same numerical answers as before because the machine is unchanged by a change in base rating.

4-5. SPECIFICATION OF TRANSFORMERS

The per unit impedance of a transformer deserves particular attention. Consider a transformer rated

2000 VA   200/400 V

for which the approximate equivalent circuit, with all impedances referred to the low-voltage side, is shown in Fig. 4-4. Let us determine the per unit

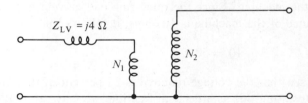

FIG. 4-4. The simplified equivalent circuit of a transformer with all series impedance transferred to the low-voltage side.

impedance of this transformer considering base quantities to be the rated values. That is, we shall determine the per unit quantities on the transformer's own base. On the low-voltage side,

base volt-amperes	2000 VA
base voltage	200 V
base current	10 A
base impedance	20 Ω

The impedance of the transformer (in pu) is

$$Z_{pu} = \frac{Z_{ohms}}{Z_{base}} = \frac{4}{20} = 0.20 \text{ pu}$$

Next let us consider the transformer with its impedance referred to the high-voltage side:

$$Z_{HV} = Z_{LV} \frac{N_2^2}{N_1^2} = Z_{LV} \frac{V_2^2}{V_1^2}$$

$$= 4 \frac{400^2}{200^2} = 16 \text{ Ω}$$

The equivalent circuit then is as shown in Fig. 4-5. Base quantities on the high-voltage side are

base volt-amperes	2000 VA
base voltage	400 V
base current	5 A
base impedance	80 Ω

The impedance of the transformer (in pu) viewed from the high-voltage side is

$$Z_{pu} = \frac{16 \text{ Ω}}{80 \text{ Ω}} = 0.20 \text{ pu}$$

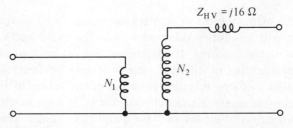

FIG. 4-5. The transformer of Fig. 4-4 with all series impedance transferred to the high-voltage side.

Hence it is seen that the per unit impedance is the *same* regardless of the side from which it is viewed. This statement (demonstrated here for a particular case) may be shown to apply generally, for *all* two-winding transformers. The equivalent circuit of the transformer is thus the simple circuit, Fig. 4-6.

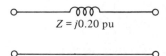

$Z = j0.20$ pu

FIG. 4-6. The simplified equivalent circuit of a transformer expressed in per unit.

4-6. SPECIFICATION OF LINES AND CABLES

Using known wire size and spacing the designer of an overhead line and cable circuit is able to calculate the ohms impedance of the circuit. To convert this impedance to per unit impedance it is necessary to determine base ohms for the circuit. This is done by assuming an arbitrary base volt-ampere rating and an arbitrary base voltage rating. The volt-ampere rating may be that assigned to other pieces of equipment on the system. The voltage rating is usually the nominal voltage of the line as determined by the voltage rating of the transformer supplying the circuit. The base impedance may then be determined from Eqs. (4-1) and (4-2). With line impedance (in ohms) calculated, the per unit impedance of the line may be determined from Eq. (4-3) as

$$Z_{pu} = \frac{Z_{ohms}}{Z_{base}} = Z_{ohms}\frac{VA_{base}}{V^2_{base}} \tag{4-7}$$

4-7. CHANGE OF BASE

Ordinarily the manufacturer of a machine expresses its characteristics in terms of per unit quantities using as a base the rated volt-amperes and rated voltage of the machine. If calculations are to be made on a system involving two machines of different ratings, it may be desirable to express the characteristics of one machine in terms of the rating of the other, or it may be desirable to express the characteristics of each machine in terms of some arbitrarily chosen set of base quantities. This is readily accomplished by remembering that the actual impedance of a machine depends only on the materials and construction, and is unchanged by a change in the rating of the machine. However, if the base is changed, the per unit impedance of the machine takes on a new value.

Suppose that a manufacturer states that a machine has $Z_{1\,pu}$ per unit impedance based on its own volt-ampere and voltage rating, which we shall define here as system 1 and in which we shall represent base quantities as VA_{B1} and V_{B1}. The question may be asked: What is the per unit impedance of this machine $Z_{2\,pu}$ in a new base system 2, in which base quantities are defined as VA_{B2} and V_{B2}?

In system 1, from Eq. (4-3),

$$Z_{\text{ohms}} = Z_{1\,pu}Z_{B1} = Z_{1\,pu}\frac{V_{B1}}{I_{B1}} = Z_{1\,pu}\frac{V_{B1}}{VA_{B1}/V_{B1}}$$

$$= Z_{1\,pu}\frac{V_{B1}^2}{VA_{B1}}$$

Similarly, in system 2,

$$Z_{\text{ohms}} = Z_{2\,pu}\frac{V_{B2}^2}{VA_{B2}}$$

Since the actual ohms of the machine are unchanged by a change of base, we can equate these two expressions for $Z_{(\text{ohms})}$ to obtain

$$Z_{1\,pu}\frac{V_{B1}^2}{VA_{B1}} = Z_{2\,pu}\frac{V_{B2}^2}{VA_{B2}}$$

Solving for $Z_{2\,pu}$ we obtain

$$Z_{2\,pu} = Z_{1\,pu}\frac{V_{B1}^2}{V_{B2}^2}\frac{VA_{B2}}{VA_{B1}} \tag{4-8}$$

This equation provides us with a means for expressing the per unit impedance of a machine in one base when its per unit impedance has already been expressed on a different base.

4-8. AN EXAMPLE SYSTEM

To illustrate the use of the per unit method in the analysis of system problems, consider the assembly of components shown in Fig. 4-7. Note that machines 1 and 2 are both rated 250 volts, as is the primary of transformer A. If this is taken as the base voltage for this part of the system, the line has a base voltage of 800 volts as determined by the turns ratio of transformer A. Transformer B was obviously not designed for this particular

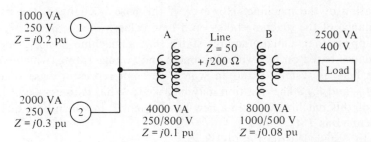

FIG. 4-7. An example power system with the characteristics of each component expressed independently.

installation, as it is rated at a higher voltage than necessary. However, its turns ratio is proper to supply the load at its rated voltage, 400 volts.

To solve a problem involving this system by the per unit method, it is necessary to determine the impedance of all components on the same base. The base selected is arbitrary. For illustration, let 5000 VA be chosen as the base and determine the per unit impedance of each part. The determination for the machines and transformers may be made using Eq. (4-8), while that for the line may be made by Eq. (4-7).

Machine 1

$$Z_{pu(5000)} = j0.2 \frac{250^2}{250^2} \frac{5000}{1000} = j1.0 \text{ pu}$$

Machine 2

$$Z_{pu(5000)} = j0.3 \frac{250^2}{250^2} \frac{5000}{2000} = j0.75 \text{ pu}$$

Transformer A

$$Z_{pu(5000)} = j0.1 \frac{250^2}{250^2} \frac{5000}{4000} = j0.125 \text{ pu}$$

Line

$$Z_{pu(5000)} = (50 + j200) \frac{5000}{800^2} = (0.39 + j1.56) \text{ pu}$$

Transformer B

$$Z_{pu(5000)} = j0.08 \frac{1000^2}{800^2} \frac{5000}{8000} = j0.078 \text{ pu}$$

Load

$$VA_{pu(5000)} = \frac{2500}{5000} = 0.5 \text{ pu}$$

The entire system may now be represented by the impedance diagram of Fig. 4-8, in which *all* components are represented on a common 5000-VA base. It is seen that the system is now a simple combination of several basic power circuits, which may be further simplified if desired.

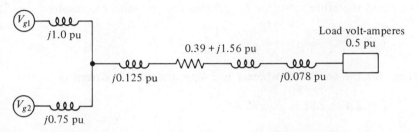

FIG. 4-8. The example system of Fig. 4-7 with the impedance of each component expressed on a 5000-VA base.

4-9. THREE-PHASE SYSTEMS; PER UNIT ANALYSIS

The previous discussion of per unit quantities was limited to single-phase machines and single-phase lines. This analysis may be readily extended to three-phase systems. Referring to the discussion on the single-phase equivalent of a three-phase system (Section 3-8), it was shown that all three-phase problems involving balanced systems may be worked on a single-phase circuit basis. In the single-phase equivalent circuit, all values are correct except the value of current, which is higher by the factor $\sqrt{3}$. From this we conclude that *any* problem involving balanced three-phase systems may be worked in terms of per unit quantities on a single-phase basis. When the solutions obtained in terms of per unit volts, per unit amperes, per unit volt-amperes, and per unit ohms are put into volts, amperes, volt-amperes, and ohms, all values are correct except the currents. The currents calculated from the single-phase equivalent diagram may then be reduced by dividing by $\sqrt{3}$ to obtain the correct numerical values applying in the actual three-phase system.

Suppose, for example, that in a system rated 1000 VA, 200 volts, the current is 2.4 pu. If this is a single-phase system, the base current is

$$I_B = \frac{VA_B}{V_B} = \frac{1000}{200} = 5 \text{ A}$$

and the actual current is

$$I = I_{\text{pu}}I_B = 5 \times 2.4 = 12 \text{ A}$$

If this is a three-phase system, the actual current is $12/\sqrt{3} = 6.95$ A.

This problem may also be approached by considering that in a three-phase system, the expression for volt-amperes is

$$VA = \sqrt{3}\,V_{LL}I_L \tag{2-113}$$

We could therefore solve for I_{base} in this relation with the result

$$I_{base\ 3\phi} = \frac{VA}{\sqrt{3}\,V} = \frac{1000}{\sqrt{3}\ \times\ 200} = 2.88\ A$$

Then, for the condition where I is 2.4 pu, the actual current is

$$I = 2.4 \times 2.88\ A = 6.95\ A$$

4-10. BASE QUANTITIES IN TERMS OF KILOVOLTS AND KILOVOLT-AMPERES

The subject of per unit quantities has been presented so far in terms of volt-amperes, volts, amperes, and ohms. In power-system practice it is common to refer to ratings in terms of kilovolt-amperes and kilovolts. Equations (4-1) to (4-7) are readily modified to accommodate this notation:

$$kV_B = \frac{I_B Z_B}{1000} \tag{4-9}$$

$$kVA_B = kV_B I_B \tag{4-10}$$

$$Z_{ohms} = Z_{pu} Z_B \tag{4-11}$$

$$I_A = I_{pu} I_B \tag{4-12}$$

$$kV_{kV} = V_{pu} kV_B \tag{4-13}$$

$$kVA_{kVA} = kVA_{pu} kVA_B \tag{4-14}$$

$$Z_{pu} = \frac{Z_{ohms}}{Z_B} = Z_{ohms}\frac{kVA_B}{kV_B^2 1000} \tag{4-15}$$

$$Z_{2pu} = Z_{1pu}\frac{kV_{B1}^2}{kV_{B2}^2}\frac{kVA_{B2}}{kVA_{B1}} \tag{4-16}$$

Problems

4-1. Consider a system in which base volt-amperes = 200 kVA and base voltage = 10,000 V.
    a. Calculate base current and base impedance.

   b. Express each of the following in terms of per unit values:
       12,000 volts
       90,000 watts
          600 ohms
           90 amperes
   c. Express each of the following in terms of percent values:
       250 volts
       450 kilovolt-amperes
       800 ohms
        10 amperes

4-2. Consider a system in which base impedance = 80 ohms and base current
      = 125 amperes.
      a. Calculate base kVA and base voltage.
      b. Express each of the following in terms of actual amperes, volts, ohms,
          or volt-amperes.
          $Z = 1.1$ pu
          $I = 27\%$
          $V = 0.75$ pu
          $VA = 36.7$ pu

4-3. Consider a system in which base current = 400 amperes and base voltage =
      25,000 volts.
      a. Calculate the actual voltage across a resistor of 4.2 pu Ω carrying a
          current of 2.5 pu A.
      b. A load has 80 percent voltage impressed across it. The current is 120
          percent and the power is 60 percent. Calculate the power factor.
      c. An impedance has 0.40 pu V impressed across it. The current is 1.4 pu A
          and the power consumed is 0.45 pu W. Calculate the value of R and
          X expressed in pu ohms and in actual ohms.

4-4. A transformer is rated 75,000 kVA, 345/138 kV, and 8 percent impedance.
      a. The transformer, under certain load conditions, carries 0.6 pu A in the
          high-voltage winding. What actual current does this represent? What
          current must be present in the low-voltage winding? Express this current
          in pu amperes.
      b. What is the impedance of the transformer (in ohms) referred to the
          high-voltage side? Referred to the low-voltage side?
      c. The low-voltage terminals of the transformer are short-circuited and
          0.22 pu voltage is applied to the high-voltage winding. Determine the
          high-side and the low-side currents that result. Express each in pu amperes
          and in actual amperes.
      d. What will be the IZ drop in the transformer if a current of 1.2 pu flows
          in the high-voltage winding.

4-5. A transformer is rated 5000 VA, 400/100 V, and 12 percent impedance.
      a. What is the impedance of the transformer (in ohms) referred to the high
          side? Referred to the low side?
      b. If for intermittent duty the transformer is re-rated 7500 VA, what will
          be the new percent impedance?

c. The same transformer is supplied to a customer who has ordered a
300/75-V machine. What will be its new volt-ampere and percent
impedance rating?

4-6. A transmission line has a series impedance of $Z = 10 + j40\,\Omega$. Calculate
the pu impedance of this line under the following conditions:
a. Operating voltage 600 V, base volt-amperes 40,000 VA.
b. Operating voltage 11,000 V, base volt-amperes 5,000,000 VA.
c. Operating voltage 22 kV, base volt-amperes 50,000 kVA.

4-7.

Redraw the system showing all impedances on a 10,000-kVA base.

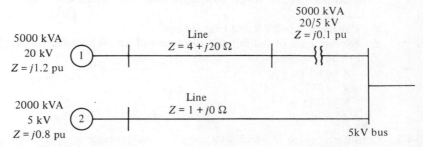

4-8. A three-phase 22-kV transmission circuit has an impedance $Z = 0.05 + j.40$
pu on a 10,000-kVA base. At the receiver end conditions are $V_r = 1.1$ pu,
$P_r = 0.4$ pu, and $f_p = 0.8$ lag. Determine V_s and I_s in terms of volts and
amperes.

chapter 5

Circuit
Constants

5-1. INTRODUCTION

The behavior of electric power systems may be accurately predicted by
calculations based on electric circuit theory. Chapter 2 presented circuit
models of many of the important components of an electric power system.
Chapter 3 showed the basis for the analysis of the behavior of these components operated as individual devices or interconnected into systems.
If numerical calculations are to be made, it is necessary to assign values
to each circuit element under consideration. The purpose of this chapter is
to provide information by which it is possible to assign correct values to
each resistance, inductance, and capacitance which appears in the circuit
models of lines, transformers, generators, loads, and other components of
electric power systems. The characteristics of overhead lines and underground cable circuits may be determined with great accuracy from the
specifications of the conductors to be used and their geometric arrangement.
The characteristics of generators and transformers may be predicted by the
designer before the machines are constructed or they may be determined by
tests after the machines have been assembled. In preliminary studies of
power-system design, it is sometimes desirable to consider machines that
have not been designed or built. For such studies it is frequently satisfactory
to use machine constants typical of a selected voltage and kilovolt-ampere
rating.

5-2. OVERHEAD LINES

The circuit model of an overhead line operating under balanced conditions of source voltage and loading was shown in Fig. 3-20. The line resistance and inductive reactance are of importance in almost all problems. For some studies it is possible to omit the capacitance and thus simplify the equivalent circuit diagram and the computations that may be made from it. This section deals with the determination of the resistance, inductive reactance, and the capacitive reactance of overhead lines based on the length of the circuit, the conductors used, and the spacing between the conductors as they are mounted on the supporting structures.

    a. Resistance. The resistance of a conductor may be calculated from the relation

$$R = \rho \frac{l}{A} \tag{2.9}$$

where ρ is the resistivity of the material, l the length of the conductor, and A the cross section of the conductor.

    In practice, several different systems of units are used in the calculation of resistance. In the MKS system, ρ is expressed in ohm-meters, length is expressed in meters, and area is expressed in square meters. Another system (CGS) expresses ρ in micro-ohm-centimeters, length in centimeters, and area in square centimeters. Another system commonly used among electric power engineers expresses resistivity in ohms circular mils per foot (sometimes designated as ohms per circular mil foot), length in feet and area in circular mils. Table A-1 (see Appendix A) summarizes two of these systems of notation and gives the value of ρ for several materials of importance in power-system work. (*Note:* A circular mil is the area of a circle having a diameter of 1 mil or 0.001 inches. The area of a circle having a diameter of 8 mils is 64 circular mils. Areas expressed in circular mils are very commonly used for conductors that have a circular cross section, such as those used in overhead lines.)

    The application of Eq. (2-9) yields the resistance of a conductor at 20°C when the value of ρ as given in Table A-1 is applied. The resistance of a conductor at any temperature may be determined by applying Eq. (2-10):

$$R_2 = R_1[1 + \alpha(T_2 - T_1)] \tag{2-10}$$

In this equation R_2 is the resistance at temperature T_2. R_1 is the resistance at temperature T_1, 20°C. Values of the temperature coefficient of resistance are given in Table A-1.

EXAMPLE 5-1. Calculate the resistance of a cylindrical aluminum conductor having the following dimensions:

length: 300 ft = 9140 cm

diameter: 0.250 in. = 250 mils = 0.635 cm

Solution. Using ρ in micro-ohm-centimeters, length must be expressed in centimeters and area in square centimeters:

$$R = \rho \frac{l}{A}$$

$$R = 2.83 \times 10^{-6} \frac{9140}{0.635^2(\pi/4)} = 0.0814 \ \Omega$$

Using ρ in ohms cir mil per foot, length must be in feet and area in circular mils:

$$R = 17.0 \frac{300}{250^2} = 0.0814 \ \Omega$$

The solution attained in Example 5-1 is for a conductor temperature T_1 of 20°C. Suppose in operation a new temperature T_2 of 130°C is observed. The resistance R_2 is

$$R_2 = R_1[1 + \alpha(T_2 - T_1)]$$
$$= 0.0814[1 + 0.0039(130 - 20)]$$
$$= 0.0814(1 + 0.429) = 0.116 \ \Omega$$

There are certain limitations in the use of Eq. (2-9) for calculating the resistance of transmission-line conductors. A slight error is introduced if the conductors are stranded rather than solid, for the individual strands are slightly longer than the length of the cable itself. When alternating current flows in a solid conductor the current does not distribute itself uniformly over the conductor cross-sectional area but tends to be of greater density near the surface of the conductor. This nonuniform distribution of current results in increased losses and causes the ac resistance of the conductor to be somewhat higher than the dc resistance. This effect is particularly noticeable in solid conductors of large cross section. It increases in importance as frequency is increased. This behavior, known as *skin effect*, causes the resistance of conductors at radio frequencies to be many times that observed with direct current. At power frequency (60 hertz) the effective resistance is only a few percent greater than that observed with dc.

In many instances, stranded conductors and composite conductors such as those shown in Fig. 5-1 are used in overhead construction practice.

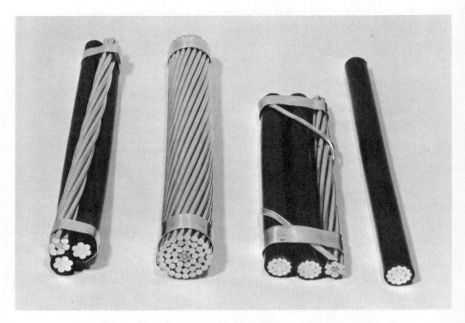

FIG. 5-1. Some typical conductors used on overhead power lines.

The composite conductor consists of a center core of high-strength material such as steel (which has a relatively high resistivity) surrounded by one or more layers of low-resistivity metal (commonly aluminum). The distribution of current between the steel and aluminum is not directly determinable by resistance calculations and so the effective resistance can only be approximated by Eq. (2-9). Advanced calculations and tests have been carried out to determine the effective resistance of all the composite conductors offered for use in overhead-line construction. These values are available in tabular form, such as Tables A-2 and A-3 in Appendix A. Tables of this sort in use in the United States usually give the resistance of conductors per thousand feet or the resistance per mile.

b. Inductance and Inductive Reactance. The physical nature of inductance may be described by reference to the single-phase line shown in cross section in Fig. 5-2. Here a current is assumed to flow out in conductor (a) and to return in conductor (b). These currents set up magnetic field lines that link between the conductors as shown. When the current changes, the flux changes and a voltage is induced in the circuit. If the current changes sinusoidally as it does in a conventional ac circuit, the voltage induced is referred to as the IX drop. As there is only a single turn in this circuit, the magnetic flux per ampere of current in the circuit is equal to the circuit inductance.

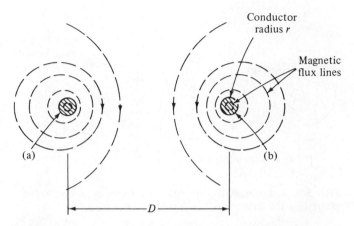

FIG. 5-2. The magnetic field associated with a single-phase line.

It is of interest to note the location of the magnetic field lines shown in Fig. 5-2. Note that most of them are in the air space between the conductors. There are, however, some magnetic field lines within the conductors themselves.

The inductance of a circuit as shown in Fig. 5-2 may be calculated from electromagnetic field relations, a subject beyond the scope of the present discussions. Assuming solid conductors carrying direct current, the inductance per meter is found to be

$$L = 2 \times 10^{-7} \left(\log_e \frac{D}{r} + \frac{1}{4} \right) \tag{5-1}$$

In this equation L is the inductance line to neutral (or the inductance per conductor) in henrys per meter of line length, D the spacing between the centers of the conductors, and r the radius of the conductors.

In the derivation of Eq. (5-1) the term $\frac{1}{4}$ arises from a consideration of the flux within the *solid* conductors. For many purposes it is desirable to eliminate this term by the introduction of a concept called the *geometric mean radius*. A physical interpretation of the geometric mean radius is illustrated in Fig. 5-3. Here the solid conductors (a) and (b) of Fig. 5-2 have been replaced by hollow tubes. The hollow tubes have a radius, termed the *geometric mean radius*, of value somewhat smaller than the radius r of the conductors in Fig. 5-2. Since there is no current within the hollow tubes, there is also no magnetic flux internal to these tubes. The magnetic flux *external* to the hollow tube in that small region between GMR and r contributes to the inductance of the circuit, an amount equal to that of the internal flux as shown in Fig. 5-2. The use of the concept of geometric mean radius permits revising Eq. (5-1) and rewriting it as Eq. (5-2). For

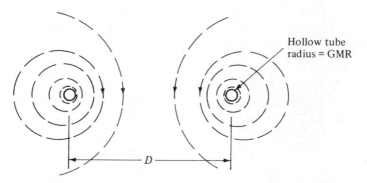

FIG. 5-3. A single-phase line with conductors of hollow tubes illustrating the meaning of Geometric Mean Radius.

solid conductors carrying direct current, the GMR is 0.778 times the physical radius:

$$L = 2 \times 10^{-7} \log_e \frac{D}{\text{GMR}} \qquad \text{henrys/meter} \qquad (5\text{-}2)$$

$$L = 0.741 \times 10^{-3} \log_{10} \frac{D}{\text{GMR}} \qquad \text{henrys/mile} \qquad (5\text{-}3)$$

With the inductance known, we may apply Eq. (2-65),

$$X_L = 2\pi f L \qquad (2\text{-}65)$$

and write an expression for the inductive reactance line to neutral (or per conductor):

$$X_L = 2\pi f\, 0.741 \times 10^{-3} \log_{10} \frac{D}{\text{GMR}} \qquad \text{ohms/mile} \qquad (5\text{-}4)$$

The use of the geometric mean radius of conductor simplifies inductance and inductive reactance calculations. The geometric mean radii of stranded conductors and of composite conductors are difficult to determine by calculation or test. Fortunately, the geometric mean radii of the conductors commonly used in overhead-line construction have been determined and are available as shown in Tables A-2 and A-3.

       c. Capacitance and Capacitive Reactance. The physical nature of the capacitance of a transmission circuit is illustrated with reference to Fig. 5-4. The two conductors, (*a*) and (*b*), similar to those of Fig. 5-2, are separated by a dielectric and hence form a capacitor. If a

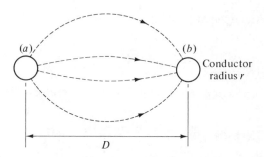

FIG. 5-4. The electric field associated with a single-phase line.

positive voltage $+V$ is put on (a) while a negative voltage $-V$ is put on (b), a charge $+q$ appears on (a) and a charge $-q$ appears on (b). Electric-field lines extend from (a) to (b) as shown in the diagram. If the voltage V changes in magnitude, the value of q must also change. Hence, during the time the voltage is changing, a current i flows lengthwise in the conductors from the source to bring about the change in the charge q on the conductor surfaces. The capacitance of this circuit may be calculated by a consideration of electromagnetic fields (a subject beyond the scope of the present discussion). Expressions for the capacitance are as follows:

$$C = \frac{2\pi\epsilon}{\log_e (D/r)} \quad \text{farad/meter} \quad \epsilon = \frac{10^{-9}}{36\pi} \quad \text{for air} \tag{5-5}$$

$$C = \frac{0.0388}{\log_{10} (D/r)} \quad \text{microfarad/mile} \tag{5-6}$$

where C is the capacitance line to neutral (or per conductor), ϵ the permittivity of the dielectric ($10^{-9}/36\pi$ for air), D the distance between the centers of the conductors, and r the physical radius of the conductors.

The capacitance line to neutral may be better understood by reference to Fig. 5-5. Here the two conductors (a) and (b) are shown with a neutral between them. The capacitance C as given by Eqs. (5-5) and (5-6) is as shown on Fig. 5-5. The capacitance between (a) and (b) is $\frac{1}{2}C$.

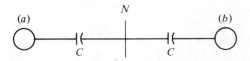

FIG. 5-5. The two conductors of a single-phase line with a neutral plane between them.

The capacitive reactance X_C is given as

$$X_C = \frac{1}{2\pi f C} \tag{2-71}$$

When this equation is combined with Eq. (5-6) there results

$$X_C = \frac{10^6}{2\pi f \times 0.0388} \log_{10} \frac{D}{r} \qquad \text{ohm–miles} \tag{5-7}$$

Note that the units given for the capacitance reactance in Eq. (5-7) are ohm-miles. As the length of the line l increases, the capacitance C increases and hence X_C decreases.

d. Constants of Single-Phase Lines. The π-line circuit model of a single-phase line, represented with a neutral, is shown in Fig. 5-6.

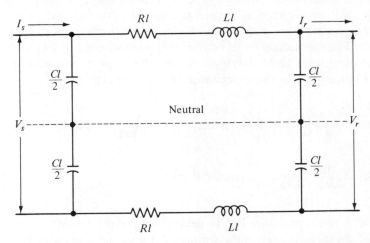

FIG. 5-6. A line-to-neutral model of a single-phase line of length l showing resistance, inductance, and capacitance.

In this diagram R, L, and C represent the resistance, inductance, and capacitance of the line per unit length and l is the line length. The values of these constants may be determined by Eqs. (2-9), (5-3), and (5-6). Note that half of the total line capacitance is placed at each end. For the solutions to problems involving sinusoidal currents and voltages, L and C must be replaced by X_L and X_C, resulting in the diagram shown in Fig. 5-7. X_L and X_C are values for unit length of line.

A study of Fig. 5-7 shows that the A to N section of this circuit behaves exactly like the B to N section, except that the voltages and currents are

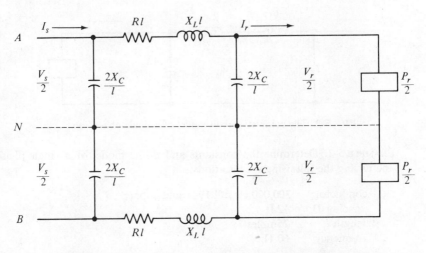

FIG. 5-7. A line-to-neutral model of a single phase line of length l, showing resistance, inductive reactance, and capacitive reactance.

180° displaced from each other in the two sections. It is therefore quite sufficient to study only a single line-to-neutral diagram, as shown in Fig. 5-8.

For many problems, particularly with short lines, the capacitive reactances at the ends of the lines may be omitted. When this is done the circuit reduces to the general power circuit discussed in detail in Chapter 3.

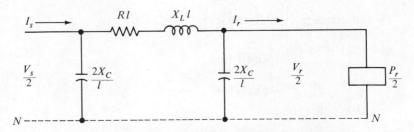

FIG. 5-8. A line-to-neutral model of one leg of a single-phase line. This may be used as the line-to-neutral model of a 3-phase line or the single-phase equivalent of a 3-phase line.

For some purposes it may be desirable to handle a single-phase line on a total circuit basis rather than on a line-to-neutral basis. For this purpose the diagram of Fig. 5-7 may be modified by the simple omission of the neutral conductor, which of course may not be present physically. The diagram may then be simplified to that of Fig. 5-9. Note that in this diagram all voltages and impedances are *twice* the values of those shown in Fig. 5-8.

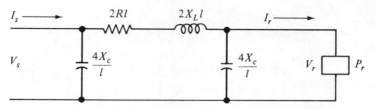

FIG. 5-9. The complete circuit model of a single-phase line.

EXAMPLE 5-2. Determine the constants and π-line model of a single-phase line having the following specifications:

conductor	300,000 cir mil 19-strand copper
spacing D	12 ft
length l	25 miles
frequency	60 Hz

Solution. From Table A-2 it may be seen that this conductor has the following characteristics:

outside diameter	0.629 in. ($r = 0.0262$ ft)
geometric mean radius	0.01987 ft
resistance per mile	0.1966 Ω (at 25°C)

R_t, total resistance (per conductor) is $R \times l$, or

$$R_t = 0.1966 \times 25 = 4.92 \ \Omega$$

The inductance (per conductor) is

$$L = 0.741 \times 10^{-3} \log_{10} \frac{D}{\text{GMR}} \qquad (5\text{-}3)$$

$$= 0.741 \times 10^{-3} \log_{10} \frac{12}{0.01987}$$

$$= 2.09 \times 10^{-3} \ \text{H/mile}$$

The total inductance per conductor is, therefore,

$$L_t = L \times l = (2.09 \times 10^{-3})(25)$$

$$= 52.2 \times 10^{-3} \ \text{H}$$

The capacitance (per conductor) is

$$C = \frac{0.0388}{\log_{10}(D/r)} \qquad (5\text{-}6)$$

$$= \frac{0.0388}{\log(12/0.0262)}$$

$$= 0.0146 \ \mu\text{F/mile}$$

The total capacitance per conductor is

$$C_t = C \times l = 0.0146 \times 25 = 0.364 \ \mu F$$

The π-line circuit model (line to neutral) with half of the capacitance on each end is as shown in Fig. 5-10. If the line operates at 60 Hz, the impedances are

$$X_L = 2\pi f L = 2\pi 60 \times 52.2 \times 10^{-3} = 19.6 \Omega$$

$$X_C = \frac{1}{2\pi f C} = \frac{10^6}{2\pi 60 \times 0.364/2} = 14{,}600 \Omega$$

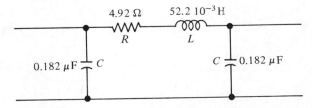

FIG. 5-10. π-line circuit model showing resistance, inductance, and capacitance, line-to-neutral.

and the line may be represented (line to neutral) as in Fig. 5-11. On a total line basis, these values are all increased by the factor 2, and the circuit becomes as shown in Fig. 5-12.

e. Constants of Three-Phase Lines. The equations for calculating the resistance, inductive reactance, and capacitive reactance of

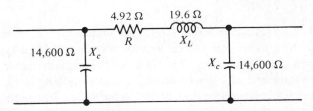

FIG. 5-11. 60-Hertz circuit model of Fig. 5-10.

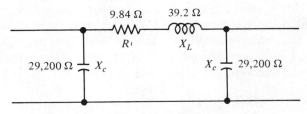

FIG. 5-12. Total circuit model.

a single-phase line on a line-to-neutral (or per conductor) basis apply equally well to the determination of these quantities for a three-phase line treated on a line-to-neutral basis, as was discussed in Section 3-7. As shown in Section 3-8, the line-to-neutral circuit model may be used as the single-phase equivalent of the three-phase circuit by appropriate modifications of currents and voltages.

Equations (5-4) and (5-7) for the calculations of inductive reactance and capacitive reactance each involve the term D, the distance between conductors. In a three-phase circuit, shown in cross section in Fig. 5-13,

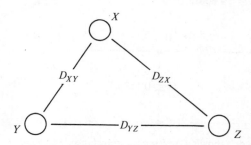
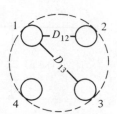

FIG. 5-13. A three-phase power line with unequal spacings between the three conductors X, Y, and Z.

FIG. 5-14. A conductor formed by a bundle of four cables.

there are three conductors, X, Y, and Z, which have unequal separation. For lines that are transposed (and with but little error for lines that are not transposed) the distance D may be determined by the relation

$$D = \sqrt[3]{D_{XY}D_{YZ}D_{ZX}} \tag{5-8}$$

On ultra-high-voltage (UHV) lines, the conductors X, Y, and Z may each be made up of several cables, as shown in Fig. 5-14, a type of construction known as *bundle conductors*. If all of these smaller cables are identical and are symmetrically spaced around a circumscribing circle as shown, it is possible to calculate an *equivalent conductor radius*. For inductive reactance calculations the equivalent geometric mean radius is given by Eq. (5-9). In this equation GMR_1 is the GMR applying to the individual cables making up the bundle. Capacitive reactance calculations are made using an equivalent radius given by Eq. (5-10), in which r_1 is the physical radius of the individual cables making up the bundle. With these two modifications, Eqs. (5-4) and (5-7) may be used to calculate the inductive reactance and the capacitance reactance per mile of line.

$$GMR_{eq} = \sqrt[n]{GMR_1 D_{12} D_{13} \cdots D_{1n}} \tag{5-9}$$

$$r_{eq} = \sqrt[n]{r_1 D_{12} D_{13} \cdots D_{1n}} \tag{5-10}$$

f. Tables of Line Constants. Data have been prepared in tabular form which make possible determination of overhead-line constants without resorting to use of equations for calculation. The construction of these tables may be explained by reference to Eqs. (5-4) and (5-7). From the well-known relation

$$\log\frac{A}{B} = \log A + \log\frac{1}{B}$$

it is possible to rewrite Eq. (5-4) as

$$X_L = 2\pi f \times 0.741 \times 10^{-3} \log_{10} D$$
$$+ 2\pi f \times 0.741 \times 10^{-3} \log_{10}\frac{1}{\text{GMR}} \qquad (5\text{-}11)$$

$$X_L = X_d + X_a \qquad (5\text{-}12)$$

We may note that (for a given frequency) the first term of Eq. (5-11) is dependent only on the spacing D, a factor determined by the geometrical arrangement of the conductors. The second term of Eq. (5-11) (for a given frequency) is dependent only on the GMR, a value which depends upon the particular conductor design selected for the line. It is therefore possible to make a tabulation (Table A-4) in which the value of

$$X_d = 2\pi f \times 0.741 \times 10^{-3} \log_{10} D \qquad (5\text{-}13)$$

is represented for each of a number of different values of D covering the range of conductor separations that might be expected in practical line construction. Other tabulations (Tables A-2 and A-3) may be prepared showing the value of

$$X_a = 2\pi f \times 0.741 \times 10^{-3} \log_{10}\frac{1}{\text{GMR}} \qquad (5\text{-}14)$$

for each of the many conductor designs available for use in construction. Now, to determine X_L for a given line design, it is only necessary to select the appropriate value of X_d from Table A-4 and to add it to the appropriate value of X_a obtained from Table A-2 or A-3.

The preparation of tables for the determination of capacitive reactance may be explained by rewriting Eqs. (5-7) as

$$X_c = \frac{10^6}{2\pi f \times 0.0388} \log_{10} D + \frac{10^6}{2\pi f \times 0.0388} \log_{10}\frac{1}{r} \qquad (5\text{-}15)$$

$$X_c = X_d' + X_a' \qquad (5\text{-}16)$$

$$X'_d = \frac{10^6}{2\pi f \times 0.0388} \log_{10} D \tag{5-17}$$

$$X'_a = \frac{10^6}{2\pi f \times 0.0388} \log_{10} \frac{1}{r} \tag{5-18}$$

Values of X'_d are shown in Table A-5. Values of X'_a are shown in Tables A-2 and A-3. To determine the capacitive reactance for a mile of line it is only necessary to select the appropriate value of X'_a and add to it the appropriate value of X'_d.

EXAMPLE 5-3. Determine the constants of a three-phase line having the following specifications:

conductor	900,000 cir mil ACSR
spacing	14 ft horizontally ◯ 14 ft ◯ 14 ft ◯
length l	125 miles
frequency	60 Hz

Solution. For computation purposes the spacing D is determined by

$$D = \sqrt[3]{D_{XY} D_{YZ} D_{ZX}}$$
$$= \sqrt[3]{14 \times 14 \times 28} = 17.6 \text{ ft} \tag{5-8}$$

From here on the determination of the impedance values for the circuit model could proceed by the method shown in Example 5-2. Instead, we shall make use of tabular values. From Table A-3 we obtain the following values:

resistance per mile (60 Hz) = 0.104 ohm
$$X_a = 0.393 \text{ ohms}$$
$$X'_a = 0.0898 \text{ megohms}$$

From Table A-4 we obtain for $D = 17.6$ ft, 60 Hz,

$$X_d = 0.3480 \text{ ohms} \qquad \text{(by interpolation)}$$

From Table A-5 we obtain for $D = 17.6$ ft, 60 Hz,

$$X'_d = 0.0851 \text{ megohms}$$

Then, by Eq. (5-12),

$$X_L = X_d + X_a = 0.348 + 0.393 = 0.741 \text{ ohms per mile}$$

and, by Eq. (5-16),

$$X_c = X'_d + X'_a = 0.0851 + 0.0898 = 0.1749 \text{ megohm miles}$$

The line-to-neutral circuit model may now be formed (see Fig. 5-15) exactly as was done for the single-phase line (Fig. 5-8).

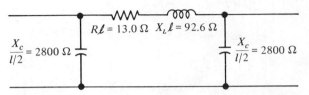

FIG. 5-15. π-line circuit model of line, Example 5-3. Line-to-neutral.

Problems pertaining to balanced three-phase loads may be solved using the circuit model of Example 5-2, as was discussed in Section 3-7. The capacitor on the right may be regarded as an addition to the load, the one on the left as an addition to the load seen by the source. For some problems the capacitors on the ends of the model may be neglected.

Figure 5-15 may also be regarded as representing the single-phase equivalent of the three-phase system. Computations may be handled as discussed in Section 3-8, with consideration being given to the capacitors as discussed in the preceding paragraph. A calculation of this type follows.

EXAMPLE 5-4. Suppose that the three-phase load on the end of the line is 12,000 Kilowatts, unity power factor, supplied at 120 kilovolts (Fig. 5-16). Considering the single-phase equivalent solution,

$$\bar{V}_r = 120,000 \,\underline{/0°}\, \text{V}$$

$$I_r = \frac{P}{V \cos \phi} = \frac{12,000,000}{120,000 \times 1} = 100 \text{ A}$$

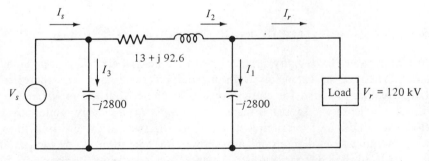

FIG. 5-16. Line of Example 5-3 carrying load. Single-phase equivalent model.

$$\bar{I}_r = 100 \;\underline{/0°}\,\text{A}$$

$$\bar{I}_1 = \frac{\bar{V}_r}{-jX_{C1}} = \frac{120{,}000 \;\underline{/0°}}{-j2800} = j42.8 \;\text{A}$$

$$\bar{I}_2 = \bar{I}_r + \bar{I}_1 = 100 + j42.8 = 109 \;\underline{/23.2°}\;\text{A}$$

$$\begin{aligned}
\bar{V}_s &= \bar{V}_r + \bar{I}_2 Z \\
&= 120{,}000 + (100 + j42.8)(13 + j92.6) \\
&= 120{,}000 - 2600 + j9820 \\
&= 117{,}400 + j9820 = 117{,}400 \;\underline{/4.8°}\;\text{V}
\end{aligned}$$

$$\bar{I}_3 = \frac{V_s}{-jX_{C3}} = \frac{117{,}400 + j9820}{-j2800} = -3.5 + j42 \;\text{A}$$

$$\begin{aligned}
\bar{I}_s &= \bar{I}_2 + \bar{I}_3 \\
&= 100 + j43 - 3.5 + j42 = 96.5 + j85 \\
&= 129 \;\underline{/41.4°}\;\text{A}
\end{aligned}$$

$$\cos \phi_s = \cos (41.4° - 4.8°) = \cos 36.6 = 0.802$$

$$\begin{aligned}
P_s = V_s I_s \cos \phi_s &= 117{,}400 \times 129 \times 0.802 \\
&= 12{,}150 \;\text{kW}
\end{aligned}$$

In the actual three-phase system (see Section 3-8),

$$I_r = \frac{100}{\sqrt{3}} = 57.7 \;\text{A}$$

$$I_s = \frac{129}{\sqrt{3}} = 74.6 \;\text{A}$$

5-3. CABLES—PORTABLE, OVERHEAD, UNDERGROUND

Insulated cables have many applications in the field of electric power. Small cables are used as extension cords around offices, homes, and factories. Larger cables are used for connections to machines that are movable over restricted distances. In some instances *portable cables* carry quite heavy electrical loads, as for example the case of electrically driven drag lines, which require several thousand horsepower for operation. *Overhead cables* find application in distribution circuits where tree conditions or proximity to buildings and other structures make the use of open-wire lines impracticable.

Underground cables are used in many situations, including major transmission circuits between large stations. Some cable power circuits operate at 345 kilovolts and carry loads of several hundred megawatts. Cables of even higher voltage will probably become available soon.

The characteristics of cables are quite different from those of overhead lines of the same voltage class. As the conductors of cables are much closer together than are those of overhead lines, the inductance and inductive reactance of cables are much smaller than similar quantities for overhead lines. In some instances the resistance of a cable may be larger in magnitude than its inductive reactance. The close spacing of conductors and the presence of solid insulation between them results in the capacitance of cables being much greater per unit length than that of overhead lines. As a consequence, the capacitive reactance of cables is much lower. For this reason care must be taken in using the short-line method of solution for problems involving cable circuits. Even though a cable is only a few miles long, many problems will demand the use of the π-line or T-line circuit model for satisfactory computation.

a. Typical Power Cables. Some typical power cables are shown in Fig. 5-17.

The conductors of insulated cables are almost always made of copper. For the same resistance per foot of length, conductors of copper are smaller than conductors of aluminum, with the result that a smaller weight of insulating material is required. Even though copper is, under present market conditions, more expensive than aluminum, the savings in the cost of insulation and the increased flexibility of smaller diameters make copper preferable to aluminum in the design of most high-voltage power cables. Underground cables using sodium as the conductor material are in limited use as distribution primary feeders.

The insulation used in power cables depends upon the application and cost. Portable cables are commonly insulated with natural or artificial rubber, a material that will stand repeated flexing even at very low temperatures. However, because of rapid deterioration of chemical structure, it cannot be used at high temperatures. Cables to be used at high temperatures may be insulated with asbestos or glass fiber.

High-voltage cables for carrying large blocks of power are frequently insulated with *oil-impregnated paper*. Other materials, such as cross-linked polyethylene, are coming into very general use for this purpose. Cables have been built using a gas, such as sulfur hexafluoride, under pressure for the insulation.

b. Calculation of Cable Constants. A *coaxial cable* consists of a center conductor (solid, stranded, or hollow) which is separated

FIG. 5-17. Some typical insulated power cables.

by insulation from a hollow cylindrical outer conductor (Fig. 5-18). The resistance to the flow of direct current may be calculated by Eq. (2-9), proper consideration being given to the area of the center conductor and the area of the outer conductor. For alternating current flow, the resistance will be higher, increasing by a few percent for 60 hertz to several hundred percent for high frequencies.

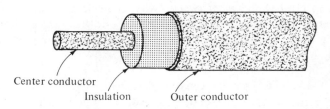

Center conductor

Insulation   Outer conductor

FIG. 5-18. A coaxial cable.

The inductance of a coaxial cable may be determined with fair accuracy by

$$L = 0.741 \times 10^{-3} \log_{10} \frac{D_o}{D_i} \quad \text{henrys/mile} \tag{5-19}$$

where D_o is the inside diameter of the outer conductor and D_i the diameter of the inner conductor. The capacitance of a coaxial cable may be determined by

$$C = \frac{0.0388k}{\log_{10}(D_o/D_i)} \quad \text{microfarads/mile} \tag{5-20}$$

where k is the relative permittivity which may have a value from 1.0 to 6.0, depending on the material of the insulation.

Coaxial cables find limited application in the electric power field. Those of the types shown in Fig. 5-17 are in common use. The constants of such cables cannot be determined by simple mathematic relations.

c. Tables of Cable Constants. The manufacturers and users of electric cables have determined the cable constants by advanced mathematical methods and by tests. Table A-6 presents the circuit constants of cables of selected designs. For reliable computations, data on cable constants for a particular design should be obtained from the manufacturer.

5-4. CHARACTERISTICS OF SYNCHRONOUS MACHINES

When a synchronous machine is being designed, studies are made which predict the machine's characteristics. After it is built, the machine is tested to determine conformance with the design expectations. From both design and test data, the manufacturer can supply very accurate data on the operating characteristics of the machine. This information can then be used to supply the constants needed in the electric-circuit models of the machine (Section 2-8a).

Engineers concerned with the design and operation of synchronous machines have made studies of the characteristics of machines of different sizes and have found that machine characteristics follow certain patterns dependent on machine kilovolt-ampere rating, voltage, speed, and field construction. Table A-7 lists some of the machine characteristics, arranged according to basic machine features.

In listing typical machine constants, it may be noted that the values given are in percent or per unit rather than in ohms, either of which may be used with the circuit model (Fig. 2-46). The use of percent or per unit quantities in expressing machine characteristics makes possible a great simplification in the tabulation of characteristics.

Each machine constructed has a nameplate rating of voltage and kilovolt-ampere capacity, which form the base quantities for the per unit system applied to each individual machine.

5-5. TRANSFORMER CHARACTERISTICS

The impedance of transformers, when expressed in terms of per unit on their own base, fall in the range from 0.01 to 0.30 per unit. When transformers are classified in terms of voltage and kilovolt-ampere rating, their per unit impedances fall into even narrower ranges. The per unit impedance of almost any single phase transformer may be determined approximately from Table A-8 and Fig. 5-19. From the table it may be noted that smaller low-voltage transformers have somewhat lower per unit impedance then larger high-voltage transformers.

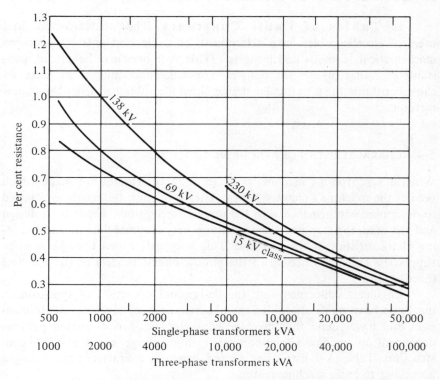

FIG. 5-19. Percent resistance of transformers based on kVA ratings.

PROBLEMS

5-1. Calculate the resistance of a bar of copper whose cross section is a square 4 mm on a side and whose length is 120 m at 20°C.

5-2. The diameter of No. 2 copper wire is 0.258 in. (Table A-2). Calculate the resistance of 2 miles of this wire
    a. at 20°C.
    b. at −50°F.
    c. at 120°F.

5-3. A transmission-line conductor is 556,500 circular mil ACSR, 105 miles in length. What is its dc resistance at 25°C small currents, at 50°C large currents? Repeat for 60-Hz conditions. What is the percentage increase?

5-4. A No. 4 solid copper wire has a diameter of 0.204 in. (Table A-2). Calculate the diameter of an aluminum wire having the same resistance per unit of length.

5-5. A single-phase 2-mile line is contructed of No. 2 solid copper wire spaced 5 ft between centers. Solve for the inductance and the 60-Hz inductive reactance per conductor using Eq. (5-1).

5-6. Determine the GMR of a No. 2 solid conductor. Repeat Problem 5-5 using Eq. (5-4).

5-7. Repeat Problem 5-5 using Eq. (5-12) and taking values of X_d and X_a from the tables in Appendix A.

5-8. A 105-mile three-phase 60-Hz 138-kV transmission line is constructed of 556,500 circular mil ACSR conductor spaced 17.5 ft horizontally. Calculate
    a. the equivalent spacing.
    b. the 60-Hz inductive reactance per conductor.

5-9. Refer to Problem 5-5. Solve for the capacitance and the 60-Hz capacitive reactance per conductor using Eqs. (5-6) and (5-7).

5-10. Repeat Problem 5-9 using Eq. (5-16) and taking values of X_d' and X_a' from the tables.

5-11. A 105-mile three-phase 60-Hz transmission line is constructed of 556,500 circular mil ACSR conductor spaced 17.5 ft horizontally. Determine the capacitive reactance per conductor.

5-12. From the results of Problems 5-2, 5-7, and 5-10, construct the π-line equivalent circuit (line to neutral Fig. 5-7) for the single-phase line. Also, construct the total circuit equivalent, Fig. 5-9.

5-13. From the results of Problems 5-3, 5-8, and 5-11, construct the π-line representation (Fig. 5-8) of the 105-mile three-phase line.

5-14.   On the basis of the thermal limitations (only) as indicated by "Approximate Current-Carrying Capacity," what maximum kVA load could be carried on the 105-mile line?

5-15.   Assume that the maximum kVA load at unity power factor is carried on the 105-mile line. What sending-end voltage is needed to provide 138 kV at the receiver? Neglect the line capacitance.

5-16.   Repeat Problem 5-15, considering the line capacitance.

5-17.   How much change in the 60-Hz inductive reactance per mile of line is occasioned by a change in wire size from No. 6 AWG to 1,000,000 circular mil, the line spacing assumed to remain constant?

5-18.   How much change in the 60-Hz inductive reactance per mile of line is occasioned by a change of spacing from 4 to 40 ft, assuming the same conductor in all cases?

5-19.   Repeat Problem 5-17, with reference to capacitive reactance.

5-20.   Repeat Problem 5-18, with reference to capacitive reactance.

5-21.   A 15-kV 60-Hz overhead line is constructed of 4/0 copper conductor spaced 4 ft horizontally. Compare the line constants of this overhead line with those of a three-conductor, belted, paper-insulated cable of similar conductor size and voltage rating.

5-22.   Determine the R and X (in percent, in per unit, and in ohms) of transformers rated as follows:

3 kVA	2500/240 V
100 kVA	2300/240/120 V
100 kVA	69/22 kV
50,000 kVA	230/138 kV class, oil immersed, self-cooled
50,000 kVA	230/69 kV class, oil immersed, forced oil cooled with forced air cooler

5-23.   Determine the synchronous impedance (percent, per unit, and ohms) of a 60,000-kVA 20-kV 3600-rpm turbine generator.

chapter 6

Assemblies of
Power-System
Components

6-1. INTRODUCTION

In previous chapters, equivalent circuits of many power-system components were presented and methods described for determining numerical values of the circuit constants that must be assigned for computational purposes. Using the per unit system, it was demonstrated that the constants of different components could be reduced to a unified system of representation. From an examination of the equivalent circuits of the various power-system components, it is obvious that each possesses certain behavior characteristics similar to those described for the basic power circuit. The performance of the basic power circuit was analyzed briefly, by graphical methods, using a transmission diagram based on the phasor representation of voltage and current in the basic power circuit (Chapter 3).

In this chapter we shall look at assemblies of power-system components which on closer examination are seen to be more complicated than the basic power circuit. However, familarity with the behavior of the basic power circuit substantially aids in understanding the behavior of the more complicated components operating individually or in combination with other devices. The practicing power engineer may observe that some of the material presented in this chapter seldom finds direct application in his daily activities. The material of this chapter is presented primarily because of its value in developing an understanding of power-system behavior.

6-2. FOUR-TERMINAL NETWORKS—SPECIAL CASE

A four-terminal network is any assembly of electrical components having two terminals defined as the *input* terminals and two other terminals, defined as the *output* terminals (Fig. 6-1a). In this section we confine our

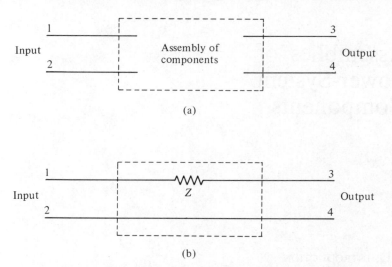

(a)

(b)

FIG. 6-1. Four terminal networks. (a) A general circuit. Internal connections are not shown. (b) The basic power circuit.

attention to the very simple form of the four-terminal network shown in Fig. 6-1b. It may be noted that in this (special) case, there is within the network no impedance terminating on the conductor, 2-4. It may be recognized at once that the general power circuit is a four-terminal network of this special class.

a. Series, Parallel, and Series–Parallel Connections of Networks. Figure 6-2 shows two four-terminal networks, 1,2,3,4 and 5,6,7,8, connected in simple series (tandem) arrangement. It is obvious that this combination may be reduced to a single four-terminal network containing a single impedance of magnitude Z_a plus Z_b.

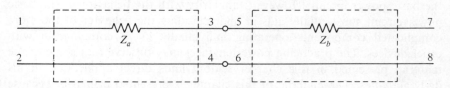

FIG. 6-2. Two simple 4-terminal networks connected in tandem.

Figure 6-3 shows two four-terminal networks 1,2,3,4 and 5,6,7,8 connected in parallel. Again it is obvious that this combination may be reduced to a single network in which there appears an impedance equal to the parallel combination of Z_a and Z_b.

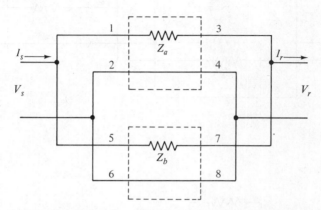

FIG. 6-3. Two simple 4-terminal networks connected in parallel.

Figure 6-4a shows two four-terminal networks, 1,2,3,4 and 5,6,7,8, connected in parallel, the combination being in series with another four-terminal network 9,10,11,12. This combination of networks may be represented in a simpler fashion as shown in Fig. 6-4b. The series–parallel combination of impedances Z_a, Z_b, and Z_c may be reduced to a single equivalent impedance (Fig. 6-1b), which is then the simplest version of the basic power circuit.

b. Networks of Lines from Source to Receiver. A more complicated form of a four-terminal network is shown in Fig. 6-5. Here it may be noted that an extensive array of impedances forms a connection between terminals 1 and 3. Note that again there are no impedances connected to line 2–4. Such an array might be a line-to-neutral representation or an equivalent single-phase representation of a three-phase system in which a number of interconnected overhead lines run from a single source 1–2 to a single receiver 3–4. A representation of this sort implies that the capacitance of the lines is being neglected.

By successive reductions the complicated network of Fig. 6-5 may be reduced to the form of the basic power circuit. The steps of such a reduction are shown by reference to Fig. 6-6. By inspection of Fig. 6-5 it may be noted that between the points marked X, Y, and Z there exists a set of three impedances, Z_e, Z_b, and Z_d, which are connected in wye. By means of a wye–delta transformation, Eq. (2-111), these impedances may be replaced by a delta connection of three other impedances, Z_k, Z_l, and Z_m, shown in Fig.

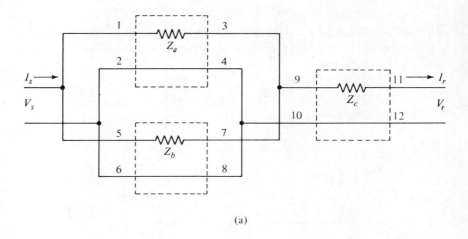

(a)

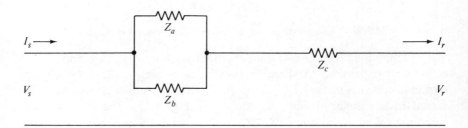

(b)

FIG. 6-4. Two simple 4-terminal networks connected in parallel, the combination in tandem with a third simple network. (a) The complete system. (b) A simplified circuit.

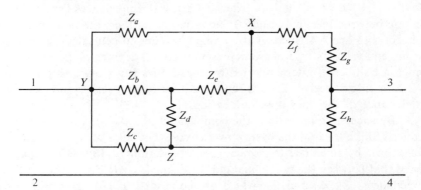

FIG. 6-5. A complicated array of impedances forming a 4-terminal network. Note that terminals 2 and 4 are directly connected.

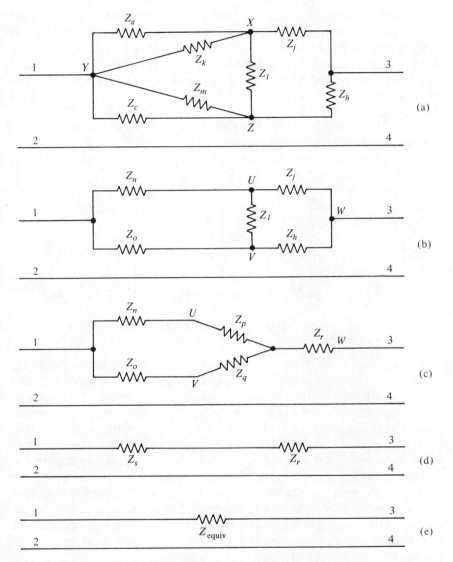

FIG. 6-6. Step-by-step reduction of the network, Fig. 6-5, to a basic power circuit.

6-6a, connected between the terminals X, Y, and Z. In this diagram it may be noted that impedances Z_a and Z_k are in simple parallel arrangement and can be replaced by Z_n, as shown in Fig. 6-6b. Similarly, impedances Z_m and Z_c of Fig. 6-6a may be replaced by their equivalent Z_o, as shown in Fig. 6-6b.

On inspection of Fig. 6-6b it may be seen that between the terminals U, V, and W there is a delta arrangement of three impedances, Z_j, Z_l, and

Z_h. A delta–wye transformation of these impedances, Eq. (2-110), results in a new diagram, as shown in Fig. 6-6c. Here it may be noted that Z_n in series with Z_p is in parallel combination with Z_o in series with Z_q. Reducing these to their equivalent Z_s yields Fig. 6-6d. This in turn may be reduced to the final equivalent circuit shown in Fig. 6-6e, which is now the basic power circuit.

c. Line with Distributed Loads—The Ladder Network. In distribution circuit practice, there sometimes occurs a *ladder-circuit arrangement* similar to that shown in Fig. 6-7. This diagram represents

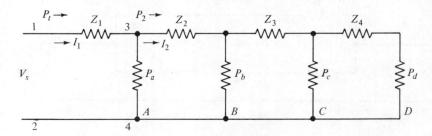

FIG. 6-7. A ladder network similar to a distribution circuit feeding several loads.

a long line which has loads connected at several points spaced randomly along its length. This arrangement may be regarded as a series of basic power circuits each in turn having as its load another circuit. For example, between terminals 1,2,3,4 we observe a single impedance Z_1 forming a basic power circuit. The load on this circuit is the entire system to the right of terminals 3-4. A problem of practical importance may be illustrated by means of this network. Suppose that at the source station the voltage V_s is known. Suppose also that from station instruments the total power output and its power factor are known. Suppose further that at each load point we have knowledge of the power and power factor. As one of the important responsibilities of distribution-system operation, it may be desirable to calculate the supply voltage at each load point. This can be done in a series of steps.

Let V_s be used as reference. I_1 may be determined in terms of magnitude and angle by the relations

$$\bar{I}_1 = \frac{P_t}{V_s F_{pt}} \underline{/\theta_t} \qquad \theta_t = \cos^{-1} F_{pt}$$

The voltage at point A may then be calculated by the relation

$$\bar{V}_a = \bar{V}_s - \bar{I}_1 \bar{Z}_1 = \bar{V}_a \underline{/\beta_a}$$

The magnitude and angle of the current supplied to the load at A may be calculated by the relations

$$\bar{I}_a = \frac{P_a}{V_a F_{pa}} \underline{/\theta_a + \beta_a} \qquad \theta_a = \cos^{-1} F_{pa}$$

The current I_2 and the voltage at load B may then be calculated by the relations

$$\bar{I}_2 = \bar{I}_1 - \bar{I}_a$$

$$\bar{V}_b = \bar{V}_a - \bar{I}_2 \bar{Z}_2$$

The procedure may be repeated until voltages at all load points have been determined.

The losses in the system may be approximated by reference to Eq. (6-1):

$$P_t = P_a + P_b + P_c + P_d + \text{losses} \qquad (6\text{-}1)$$

A determination of greater accuracy might be made by calculating the losses in each line section from known currents and circuit resistances.

It is obvious that this problem could also be worked by beginning at the extreme load end of the line if the conditions there were known or assumed. A calculation proceeding through this system from right to left would then give information as to the required conditions at the source station.

EXAMPLE 6-1. Determine voltage condition on a system of the type shown in Fig. 6-7, assuming that

$V_s = 5000$ V

$P_t = 600$ kW   0.8 power-factor lag

$Z_1 = 4 \underline{/70°}\ \Omega$

$P_2 = 400$ kW   0.707 power-factor lag

$Z_2 = 2.5 \underline{/60°}\ \Omega$

Solution.
a. At the source,

$$I_1 = \frac{P_t}{V_s \cos \varphi_t} = \frac{600{,}000}{5000 \times 0.8} = 150 \text{ A}$$

Considering V_s as reference,

$$\bar{V}_s = 500 \; \underline{/0°} \; \text{V}$$

$$I_1 = 150 \; \underline{/-37°} \; \text{A}$$

$$I_1 Z_1 = 150 \; \underline{/-37°} \times 4 \; \underline{/70°} = 600 \; \underline{/33°} \; \text{V}$$

$$\begin{aligned}
\bar{V}_A = \bar{V}_S - I_1 Z_1 &= 5000 \; \underline{/0°} - 600 \; \underline{/33°} \\
&= 5000 + j0 - (503 + j326) \\
&= 4497 - j326 \\
&= 4500 \; \underline{/-4.1°} \; \text{V}
\end{aligned}$$

b. At A,

$$I_2 = \frac{P_2}{V_A \cos \varphi_2} = \frac{400,000}{4500 \times 0.707} = 125 \; \text{A}$$

Since $\bar{V}_A = 4500 \; \underline{/-4.1°} \; \text{V}$,

$$I_2 = 125 \; \underline{/-45° - 4.1°} = 125 \; \underline{/-49°} \; \text{A}$$

$$I_2 Z_2 = 125 \; \underline{/-49°} \times 2.5 \; \underline{/60°} = 312 \; \underline{/11°} \; \text{V}$$

$$\begin{aligned}
\bar{V}_B = \bar{V}_A - I_2 Z_2 &= 4497 - j326 - (308 + j60) \\
&= 4189 - j386 \\
&= 4190 \; \underline{/-5.2°} \; \text{V}
\end{aligned}$$

This method of solution could be carried on to all sections of the line if desired.

6-3. FOUR-TERMINAL NETWORKS; A, B, C, D CONSTANTS

a. General Four-Terminal Network. The general four-terminal network is described with the aid of Fig. 6-8. This network is described by two source-terminals between which there is a voltage V_s and through which a current I_s flows, as shown in the diagram. There are also

FIG. 6-8. The general 4-terminal network consisting of passive impedances connected in any fashion.

two receiver-terminals between which the voltage is V_r and through which the current I_r flows. The network consists of any arrangement of electrical components. We shall limit this discussion to any array in which all are passive elements; that is, there are no current or voltage sources within the network. Unlike the simple network treated in Section 6-2, the arrangement of components may be such that the current I_s is quite different from the current I_r.

From general network theory (not presented here) it can be shown that, at the terminals, the behavior of the network may be specified by equations having the form

$$\bar{V}_s = A\bar{V}_r + B\bar{I}_r \tag{6-2}$$

$$\bar{I}_s = \bar{V}_r C_r + D\bar{I}_r \tag{6-3}$$

$$AD - BC = 1 \tag{6-4}$$

In general, all terms in these equations are *complex numbers*. For a given network, A, B, C, and D are constants. From Eqs. (6-2) and (6-3) it may concluded that if values are assigned to any two of the terms V_s, V_r, I_s and I_r, the other two of the four terms are uniquely determined. Equation (6-4) shows that A, B, C, and D are themselves interrelated.

It is the purpose of this section to analyze the nature of four-terminal networks and to become familiar with their characteristics. In many instances a complicated four-terminal network may, on examination, be found to be an assembly of simpler four-terminal networks. In this situation it is possible to approach the determination of the values of A, B, C, and D for the complicated network by manipulating the constants applicable to the several simpler networks forming the whole. We shall, therefore, determine the constants of several of the simple basic networks and then examine some of the methods by which they may be put into combination to form more complicated systems.

b. Series Impedance. Our study of the basic power circuit containing a simple series impedance (Fig. 6-9) is, in fact, an analysis of one of the elementary forms of the four-terminal network. For the network in

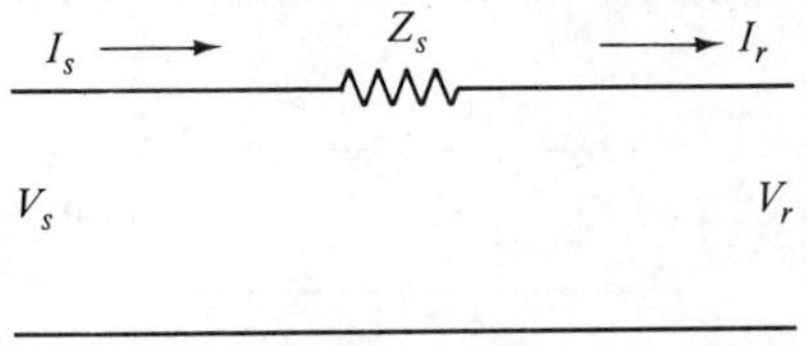

FIG. 6-9. A 4-terminal network consisting of a single series impedance.

Fig. 6-9 we may write expressions for V_s and I_s in terms of V_r and I_r as follows:

$$\bar{V}_s = \bar{V}_r + Z_s \bar{I}_r$$

$$\bar{I}_s = 0\bar{V}_r + \bar{I}_r$$

A comparison of these two equations with Eqs. (6-2) and (6-3) permits us to write the A, B, C, and D constants as follows:

$$A = 1.0 \qquad B = Z_s \qquad C = 0 \qquad D = 1.0 \tag{6-5}$$

c. Shunt Admittance. Another simple four-terminal network consists of a single shunt admittance, Y, connected as shown in Fig 6-10.

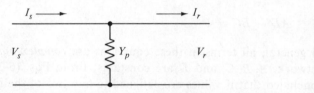

FIG. 6-10. A 4-terminal network consisting of a single shunt admittance.

For this network we may write expressions for V_s and I_s in terms of V_r and I_r as follows:

$$\bar{V}_s = \bar{V}_s + 0\bar{I}_r$$

$$\bar{I}_s = Y_p \bar{V}_r + \bar{I}_r$$

Again comparing these equations with Eqs. (6-2) and (6-3) we may declare values of A, B, C, and D as follows:

$$A = 1.0 \qquad B = 0 \qquad C = Y_p, \qquad D = 1.0 \tag{6-6}$$

d. Two Four-Terminal Networks in Tandem. Two four-terminal networks may be connected in tandem as shown in Fig. 6-11, to form a new four-terminal network. Here it is assumed that network

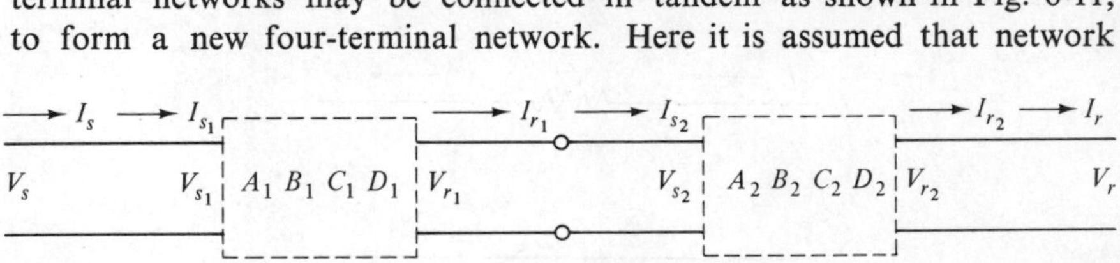

FIG. 6-11. Two general 4-terminal networks connected in tandem.

1 has the constant A_1, B_1, C_1, and D_1, while network 2 has the the constant A_2, B_2, C_2, and D_2. The two in combination form a system that has input V_s and I_s and output V_r and I_r. Considering only network 2 we may write

$$\bar{V}_{s2} = A_2\bar{V}_{r2} + B_2\bar{I}_{r2} = A_2\bar{V}_r + B_2\bar{I}_r$$

$$\bar{I}_{s2} = C_2\bar{V}_{r2} + D_2\bar{I}_{r2} = C_2\bar{V}_r + D_2\bar{I}_r$$

Considering only network 1 we may write

$$\bar{V}_{s1} = A_1\bar{V}_{r1} + B_1\bar{I}_{r1}$$

$$\bar{I}_{s1} = C_1\bar{V}_{r1} + D_1\bar{I}_{r1}$$

Noting that

$$\bar{V}_s = \bar{V}_{1s} \qquad \bar{I}_s = \bar{I}_{s1}$$

$$\bar{V}_{r1} = \bar{V}_{s2} \qquad \bar{I}_{r1} = \bar{I}_{s2}$$

$$\bar{V}_{r2} = \bar{V}_r \qquad \bar{I}_{r2} = \bar{I}_r$$

we may write that

$$\bar{V}_s = A_1(A_2\bar{V}_r + B_2\bar{I}_r) + B_1(C_2\bar{V}_r + D_2\bar{I}_r)$$

$$\bar{I}_s = C_1(A_2\bar{V}_r + B_2\bar{I}_r) + D_1(C_2\bar{V}_r + D_2\bar{I}_r)$$

Collecting terms necessary to put the expressions for V_s and I_s in standard form we have

$$\bar{V}_s = (A_1A_2 + B_1C_2)\bar{V}_r + (A_1B_2 + B_1D_2)\bar{I}_r \tag{6-7}$$

$$\bar{I}_s = (C_1A_2 + D_1C_2)\bar{V}_r + (C_1B_2 + D_1D_2)\bar{I}_r \tag{6-8}$$

these equations in turn may be rewritten as

$$\bar{V}_s = A_{eq}\bar{V}_r + B_{eq}\bar{I}_r$$

$$\bar{I}_s = C_{eq}\bar{V}_r + D_{eq}\bar{I}_r$$

where the equivalent constants are

$$A_{eq} = A_1A_2 + B_1C_2 \tag{6-9}$$

$$B_{eq} = A_1B_2 + B_1D_2 \tag{6-10}$$

$$C_{eq} = C_1 A_2 + D_1 C_2 \tag{6-11}$$

$$D_{eq} = C_1 B_2 + D_1 D_2 \tag{6-12}$$

The above equations permit us to evaluate the equivalent A, B, C, and D constants for two known networks connected in tandem. Attention must be directed to the facts that the order in which the networks are connected in tandem is of importance. If in Fig. 6-11 the two networks 1 and 2 had been interchanged in position, the equivalent constants would be different from those given in Eqs. (6-9) to (6-12).

EXAMPLE 6-2. Determine the equivalent $ABCD$ constants of network 1 connected in tandem with network 2 as indicated in Fig. 6-12.
Solution.
a. In network 1, by Eq. (6-5),

$$A_1 = 1.0 \qquad B_1 = 20\ \underline{/30°} \qquad C_1 = 0 \qquad D_1 = 1.0$$

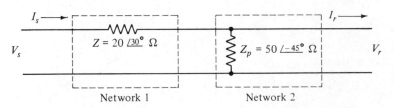

FIG. 6-12. Two networks in tandem, Example 6-2.

b. In network 2,

$$Y_P = \frac{1}{Z_P} = \frac{1}{50\ \underline{/-45°}} = 0.02\ \underline{/45°}\ \text{mho}$$

By Eq. (6-6),

$$A_2 = 1.0 \qquad B_2 = 0 \qquad C_2 = 0.02\ \underline{/45°} \qquad D_2 = 1.0$$

c. Combining the two networks, by Eq. (6-9),

$$\begin{aligned}
A_{eq} &= A_1 A_2 + B_1 C_2 \\
&= 1.0 \times 1.0 + 20\ \underline{/30°} \times 0.02\ \underline{/45°} \\
&= 1.0 + 0.4\ \underline{/75°} = 1.0 + 0.104 + j0.39 \\
&= 1.104 + j0.39 = 1.17\ \underline{/19.5°}
\end{aligned}$$

By Eq. (6-10),

$$B_{eq} = A_1B_2 + B_1D_2$$
$$= 1.0 \ \underline{/0°} \times 0 + 20 \ \underline{/30°} \times 1.0 \ \underline{/0°}$$
$$= 20 \ \underline{/30°}$$

By Eq. (6-11),

$$C_{eq} = C_1A_2 + D_1C_2$$
$$= 0 \times 1.0 + 1.0 \times 0.02 \ \underline{/45°}$$
$$= 0.02 \ \underline{/45°}$$

By Eq. (6-12),

$$D_{eq} = C_1B_2 + D_1D_2$$
$$= 0 \times 0 + 1.0 \times 1.0$$
$$= 1.0 \ \underline{/0°}$$

The same results could have been obtained by considering receiver-end conditions as V_r and I_r, then solving the ladder network for V_s and I_s in terms of V_r and I_r.

e. Two Four-Terminal Networks in Parallel. Two four-terminal networks may be connected in parallel as shown in Fig. 6-13. A new four-terminal network may be determined which will be equivalent in performance to the two networks in parallel. Note that

$$\bar{V}_{s1} = \bar{V}_{s2} = \bar{V}_s$$
$$\bar{V}_{r1} = \bar{V}_{r2} = \bar{V}_r$$

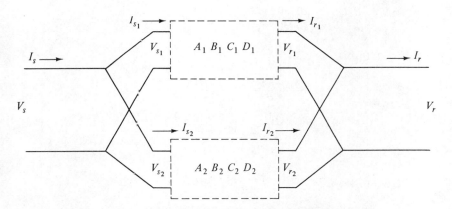

FIG. 6-13. Two general 4-terminal networks connected in parallel.

$$\bar{I}_s = \bar{I}_{s1} + \bar{I}_{s2}$$

$$\bar{I}_r = \bar{I}_{r1} + \bar{I}_{r2}$$

By a consideration of the above relations and by manipulations somewhat like those made for the tandem arrangement, it may be shown that the four-terminal network equivalent to that shown in Fig. 6-13 will have constants as follows:

$$A_{eq_p} = \frac{A_1 B_2 + A_2 B_1}{B_1 + B_2} \tag{6-13}$$

$$B_{eq_p} = \frac{B_1 B_2}{B_1 + B_2} \tag{6-14}$$

$$C_{eq_p} = C_1 + C_2 + \frac{(A_1 - A_2)(D_2 - D_1)}{B_1 + B_2} \tag{6-15}$$

$$D_{eq_p} = \frac{B_2 D_1 + B_1 D_2}{B_1 + B_2} \tag{6-16}$$

EXAMPLE 6-3. Refer to Example 6-2. Consider that network 1 is connected in parallel with network 2. Determine the equivalent $ABCD$ constants.
Solution.
By Eq. (6-13),

$$A_{eq_p} = \frac{A_1 B_2 + A_2 B_2}{B_1 + B_2} = \frac{1.0 \times 0 + 1.0 \times 20 \,\underline{/30^\circ}}{20 \,\underline{/30^\circ} + 0} = 1.0 \quad \underline{/0^\circ}$$

By Eq.(6-14),

$$B_{eq_p} = \frac{B_1 B_2}{B_1 + B_2} = \frac{20 \,\underline{/30^\circ} \times 0}{20 \,\underline{/30^\circ} + 0} = 0$$

By Eq. (6-15),

$$C_{eq_p} = C_1 + C_2 + \frac{(A_1 - A_2)(D_2 - D_1)}{B_1 + B_2}$$

$$= 0 + 0.02 \,\underline{/45^\circ} + \frac{(1.0 - 1.0)(1.0 - 1.0)}{20 \,\underline{/30^\circ} + 0}$$

$$= 0.02 \,\underline{/45^\circ}$$

By Eq. (6-16),

$$D_{eq_p} = \frac{B_2 D_1 + B_1 D_2}{B_1 + B_2} = \frac{0 \times 1.0 + 20 \,\underline{/30^\circ} \times 1.0}{20 \,\underline{/30^\circ} + 0}$$

$$= 1.0 \quad \underline{/0^\circ}$$

It may be noted that these two networks in parallel have the identical characteristics of network 2 alone.

f. More Complicated Networks. The determination of the *ABCD* constants of complicated networks may involve considerable detailed computation. If the network can be broken up into a group of simpler four-terminal networks that are in series, in parallel, or in series–parallel arrangement, the relations so far presented will permit evaluation of the constants.

For some configurations it will be necessary to resort to general network analysis and make a solution through the application of Kirchhoff's voltage law and Kirchhoff's current law. For such determinations it is frequently desirable to use digital computers. The solutions of problems of this kind are beyond the interest of the present discussion. However, it is important to note that methods are available for calculating the *ABCD* constants of any generalized four-terminal network.

g. Experimental Determination of *ABCD* Constants. If a four-terminal network has been constructed and can be tested, its *ABCD* constants may be readily determined regardless of the complexity of the network. Two tests are sufficient: one in which the receiver terminals are *open-circuited* and one in which the receiver terminals are *short-circuited*.

When the receiver terminals are open-circuited (Fig. 6-14) I_r is 0 and Eqs. (6-2) and (6-3) reduce to

FIG. 6-14. A 4-terminal network. The receiver terminals are open-circuited for test. $I_r = 0$.

$$\bar{V}_s = A\bar{V}_r$$

$$\bar{I}_s = C\bar{V}_r$$

The values of A and C may be determined from measured phasor values of V_s, I_s, and V_r.

In the second test, the receiver terminals are short-circuited (Fig. 6-15) and measurements made of the phasor values of V_s, I_s, and I_r. The constants B and D are then determinable from the relations

$$\bar{V}_s = B\bar{I}_r$$

$$\bar{I}_s = D\bar{I}_r$$

$I_s \longrightarrow$

V_s

$\longrightarrow I_r$

$V_r = 0$

FIG. 6-15. A 4-terminal network. The receiver terminals are short-circuited for test. $V_r = 0$.

EXAMPLE 6-4. Determine the $ABCD$ constants of a network on which the following test results have been observed.
a. Receiver open-circuited, $I_r = 0$:

$$\bar{V}_s = 100 \, \underline{/0^\circ} \text{ V}$$

$$\bar{V}_r = 70.7 \, \underline{/-45^\circ} \text{ V}$$

$$\bar{I}_s = 1.41 \, \underline{/-45^\circ} \text{ A}$$

b. Receiver short-circuited, $V_r = 0$:

$$\bar{V}_s = 100 \, \underline{/0^\circ} \text{ V}$$

$$\bar{I}_s = 2.0 \, \underline{/-90^\circ} \text{ A}$$

$$\bar{I}_r = 2.0 \, \underline{/-90^\circ} \text{ A}$$

Solution:
From a,

$$\bar{V}_s = A\bar{V}_r$$

$$100 \, \underline{/0^\circ} = A \times 70.7 \, \underline{/-45^\circ}$$

$$A = 1.41 \, \underline{/45^\circ}$$

and

$$\bar{I}_s = C\bar{V}_r$$

$$1.41 \, \underline{/-45^\circ} = C \times 70.7 \, \underline{/-45^\circ}$$

$$C = 0.02 \, \underline{/0^\circ}$$

From b,

$$\bar{V}_s = B\bar{I}_r$$

$$100 \, \underline{/0^\circ} = B \times 2.0 \, \underline{/-90^\circ}$$

$$B = 50 \, \underline{/90^\circ}$$

and

$$I_s = DI_r$$

$$2.0 \underline{/-90°} = D \times 2.0 \underline{/-90°}$$

$$D = 1.0 \underline{/0°}$$

h. $ABCD$ Constants of a π Circuit. The π-line representation of a transmission circuit with capacitance considered has been discussed on several previous ocassions. Other circuits arranged in this general form are frequently found in power-system practice. The common occurrence of the π circuit justifies its special treatment in terms of generalized circuit constants.

The π-network is shown in general form in Fig. 6-16. Note that the shunt branches are designated as admittances Y_1 and Y_2, the series branch as an impedance Z. This is merely a matter of convenience in notation.

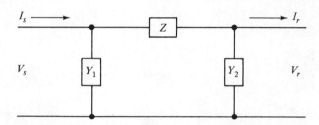

FIG. 6-16. A π-circuit 4-terminal network.

The π-circuit may be treated as the tandem connection of three four-terminal networks as shown in Fig. 6-17. The $ABCD$ constants of part (b) may be determined from Eq. (6-5); the $ABCD$ constants of parts (a) and (c) may be determined from Eq. (6-6). Knowing the $ABCD$ constants of parts (a) and (b) it is possible to combine them in tandem to form a new network by the method described in Section 6-3d. This new network may then be

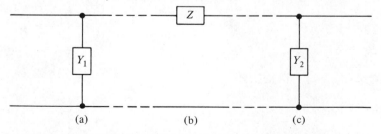

FIG. 6-17. A π-circuit treated as three 4-terminal networks connected in tandem.

combined in tandem with part (c) to give the *ABCD* constants of the total π circuit.

An alternative method for determining the *ABCD* constants of the π circuit is illustrated by means of Fig. 6-18. Here it will be assumed that

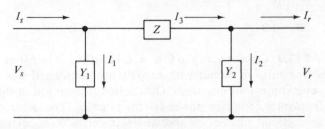

FIG. 6-18. A π-circuit labeled for the determination of *ABCD* constants.

the values of V_r and I_r are both known. From these known terminal conditions the behavior of each part of the ladder network is determined until eventually V_s and I_s have been expressed in terms of V_r and I_r. The development proceeds as follows:

$$\bar{I}_2 = \bar{V}_r Y_2$$

$$\bar{I}_3 = \bar{I}_r + \bar{I}_2 = \bar{I}_r + \bar{V}_r Y_2$$

$$\bar{V}_s = \bar{V}_r + \bar{I}_3 Z = \bar{V}_r + (\bar{I}_r + \bar{V}_r Y_2)Z$$

$$\bar{I}_s = \bar{I}_3 + \bar{I}_1 = \bar{I}_3 + \bar{V}_s Y_2 = \bar{I}_r + \bar{V}_r Y_2 + [\bar{V}_r + (\bar{I}_r + \bar{V}_r Y_2)Z]Y_1$$

Putting this into the standard form of Eqs. (6-2) and (6-3) we observe that

$$\bar{V}_s = (1 + Y_2 Z)\bar{V}_r + Z\bar{I}_r$$

$$\bar{I}_s = (Y_2 + Y_1 + Y_1 Y_2 Z)\bar{V}_r + (1 + ZY_1)\bar{I}_r$$

From these equations it follows that

$$A_\pi = 1 + Y_2 Z \tag{6-17}$$

$$B_\pi = Z \tag{6-18}$$

$$C_\pi = Y_2 + Y_1 + Y_1 Y_2 Z \tag{6-19}$$

$$D_\pi = 1 + ZY_1 \tag{6-20}$$

From these equations, the π-line representation of a transmission circuit

(Fig. 5-8) may be expressed by $ABCD$ constants by observing that

$$Z = (R + jX_L)l \tag{6-21}$$

$$Y_1 = Y_2 = j\frac{l}{2X_c} \tag{6-22}$$

where X_L and X_c are the reactances per unit length and l is the line length.

EXAMPLE 6-5. Determine the $ABCD$ constants applying to the line of Example 5-3.

Solution. In Example 5-3 the π-circuit representation was shown to be as in Fig. 6-19. Here

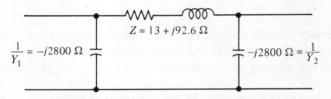

FIG. 6-19. Circuit diagram for Example 6-5.

$$Z = 13 + j92.6 = 93\ \underline{/82°}\ \Omega$$

$$Y_1 = \frac{1}{-j2800} = 0.000357\ \underline{/90°}\ \text{mho}$$

$$Y_2 = \frac{1}{-j2800} = 0.000357\ \underline{/90°}\ \text{mho}$$

Then

$$\begin{aligned}
A_\pi &= 1 + Y_2 Z = 1 + 93\ \underline{/82°} \times 0.00036\ \underline{/90°} \\
&= 1 + 0.034\ \underline{/172°} = 1 - 0.033 + j0.004 \\
&= 0.967 + j0.004 = 0.967\ \underline{/0.2°}
\end{aligned}$$

$$B_\pi = 93\ \underline{/82°}$$

$$\begin{aligned}
C_\pi &= Y_1 + Y_2 + Y_1 Y_2 Z \\
&= 0.00036\ \underline{/90°} + 0.00036\ \underline{/90°} + 0.00036\ \underline{/90°} \times 0.00036\ \underline{/90°} \times 93\ \underline{/82°} \\
&= -0.0000017 + j0.00071 \\
&= 0.000702\ \underline{/90.14°}
\end{aligned}$$

$$D_\pi = A_\pi = 0.967\ \underline{/0.2°}$$

6-4. TRANSMISSION DIAGRAMS DERIVED FROM $ABCD$ CONSTANTS

A brief discussion of transmission diagrams related to the basic power circuit was presented in Section 3-5. The diagrams presented there will be further developed in this section as related to the more general circuit arrangements that may be expressed through $ABCD$ constants.

a. Comparison of Basic Power Circuit with Generalized Circuit. The equations describing the basic power circuit are

$$\bar{V}_s = \bar{V}_r + \bar{I}_r Z$$

$$\bar{I}_s = \bar{I}_r$$

As has been previously pointed out, these equations arise from a consideration of a network consisting of a single series impedance. Under this circumstance the sending-end current is exactly equal to the receiving-end current.

The equations describing the generalized four-terminal network are

$$\bar{V}_s = A\bar{V}_r + B\bar{I}_r \tag{6-2}$$

$$\bar{I}_s = C\bar{V}_r + D\bar{I}_r \tag{6-3}$$

These more complicated equations arise when connections exist, as in power-system practice, from line to neutral. These connections appear when we consider items such as line capacitance, fixed loads within the network, and shunt reactors.

Generalized circuit constants may be developed from a consideration of impedance expressed in ohms, voltage expressed in volts, and currents expressed in amperes. If the analysis is made on this basis, term A of Eq. (6-2) must be dimensionless; term B will have dimensions of ohms because its product with current must be volts. Similarly, in Eq. (6-3) term C will have dimensions of mhos, for its product with volts must be amperes, and the term D must be dimensionless.

If the analysis of the network has been made on the basis of per unit quantities A, B, C, and D will all be dimensionless. In general, all these quantities will be complex numbers.

b. Transmission Diagram Construction. In Section 3-5, a method was outlined by which the basic power circuit could be analyzed graphically with the aid of a transmission diagram. A similar method of study may be employed with the generalized four-terminal network, although the construction of the transmission diagram is somewhat more complicated. Furthermore, it is necessary to construct a diagram pertaining to current in addition to the diagram pertaining to voltage. The construction of these diagrams will be illustrated for a circuit in which the following *ABCD* constants apply. These constants are quite typical of a power system consisting of a generator, transformer, high-voltage transmission line, and step-down transformer:

$$A = A \underline{/\alpha} = 0.80 \underline{/6°} \text{ pu}$$

$$B = B \underline{/\beta} = 0.50 \underline{/80°} \text{ pu}$$

$$C = C \underline{/\gamma} = 0.25 \underline{/70°} \text{ pu}$$

$$D = D \underline{/\delta} = 1.10 \underline{/-2°} \text{ pu}$$

The start of the construction of the voltage diagram is illustrated in Fig. 6-20. Using V_r as reference and assuming it to have a value of 1.0 $\underline{/0°}$ pu,

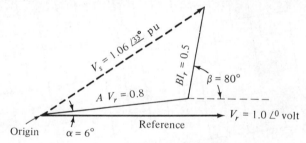

5 cm = 1.0 pu voltage

FIG. 6-20. A phasor diagram for determining V_s under the conditions that

$$\bar{V}_r = 1.0 \underline{/0°}$$

$$I_r = 1.0 \underline{/0°}$$

$$A = 0.8 \underline{/6°}$$

$$B = 0.5 \underline{/80°}$$

the phasor of V_r is laid out to scale horizontally as shown. AV_r, which now has the value 0.80 $\underline{/6°}$, is laid out on the diagram as shown. Next assume the receiver current to be 1.0 $\underline{/0°}$ pu. Now BI_r takes on the value 0.50 $\underline{/80°}$. The phasor BI_r is added to the tip of the phasor AV_r in accordance with Eq. (6-2). The phasor V_s now represents in magnitude and phase position the source end voltage under the conditions of receiver voltage being 1.0 $\underline{/0°}$ and receiver current 1.0 $\underline{/0°}$.

Following the procedure outlined in Section 3.5, phasors of BI_r may be located to correspond with I_r having the values 1.0 $\underline{/-90°}$ and 1.0 $\underline{/+90°}$, corresponding to load-power factors of 0 lag and 0 lead. The diagram may now be expanded to account for any chosen value of receiver current by the addition of the grid shown in Fig. 6-21. Since the value of AV_r is directly proportional to the magnitude of V_r, origins corresponding to different values of V_r may be designated as shown on the diagram.

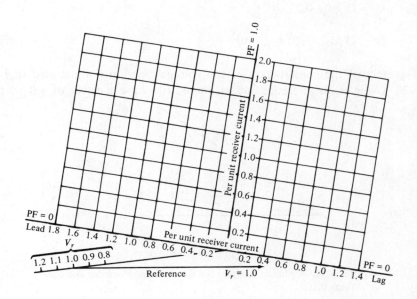

Scale: 50 mm = 1.0 pu voltage

FIG. 6-21. A voltage diagram relating V_s to V_r and I_r for the conditions

$$A = 0.8 \ \underline{/6°}$$

$$B = 0.5 \ \underline{/80°}$$

The construction of the current diagram follows a similar procedure (Fig. 6-22). Again V_r is used as reference and so is laid out horizontally. Assuming V_r to be 1.0 $\underline{/0°}$ pu, the phasor CV_r is evaluated (0.25 $\underline{/70°}$) and laid out to scale from the origin of the diagram as shown.

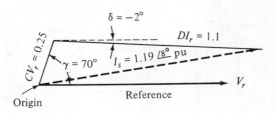

5 cm = 1 pu current

FIG. 6-22. A phasor diagram for determining I_s under the conditions that

$$\bar{V}_r = 1.0 \ \underline{/0°}$$

$$I_r = 1.0 \ \underline{/0°}$$

$$C = 0.25 \ \underline{/70°}$$

$$D = 1.1 \ \underline{/-2°}$$

Again assuming I_r to have a magnitude of 1.0 $\underline{/0°}$ pu, the phasor DI_r (1.1 $\underline{/-2°}$) is added in vector fashion to the phasor $\overline{CV}_r$, as shown. A line drawn from the origin to the tip of the DI_r phasor represents I_s in magnitude and angle on the assumption that V_r is 1.0 $\underline{/0°}$ pu and I_r is 1.0 $\underline{/0°}$ pu.

Following the method used in constructing the voltage diagram, it is now possible to draw a grid on which any value of receiver current may be represented, as shown in Fig. 6-23. Origins may be designated for different values of V_r as indicated. The diagrams (Figs. 6-21 and 6-23) may be used for the graphical solution of many problems pertaining to the four-terminal network with *ABCD* constants as specified.

6-5. GRAPHICAL ANALYSIS OF THE FOUR-TERMINAL NETWORK

The transmission diagrams (Figs. 6-21 and 6-23) may now be used to analyze some of the characteristics of the four-terminal network which are of interest to the electric-system operator.

It is very commonly recognized that electric power must be delivered to the customers in any desired amount, while voltage and frequency are held within specified limits. Those persons directly associated with system operation are aware that another quantity, reactive power (or quadrature power), has a great influence on the control of voltage and on system losses and in some systems requires the same attention in system dispatch as does the flow of power itself. Reactive power may be positive or negative. Reactive power fed to a receiver is said to be *positive* when the receiver current

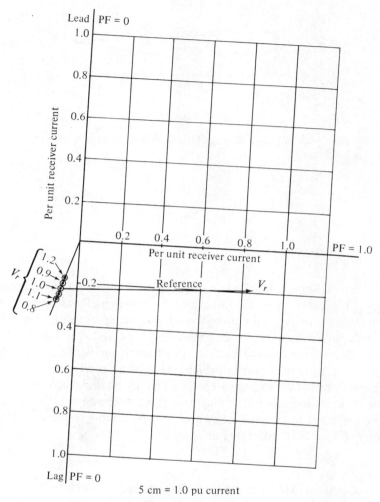

5 cm = 1.0 pu current

This scale applies only to the measurement of I_s

FIG. 6-23. A current diagram relating I_s to V_r and I_r for the conditions

$$C = 0.25 \; \underline{/70°}$$

$$D = 1.1 \; \underline{/-2°}$$

lags the receiver voltage. It is *negative* when the receiver current *leads* the voltage.

Transformers, induction motors, and all equipment having iron cores that must be magnetized by currents supplied by the power system require lagging currents and so are said to be supplied with positive reactive power. The electrical equipment found on the customers' premises are predominately of this type. Capacitors, overexcited synchronous motors, overhead transmission lines, and cables are all supplied with currents which have components leading the voltage, and so are supplied with negative reactive power. In the analysis of system behavior which follows, attention should be directed to the effects on system response of both real power and reactive power.

a. **Receiver Voltage Held Constant.** In many respects it is desirable to hold the voltage constant at the receiver end of a power circuit. To do so under varying load conditions requires appropriate changes of voltage at the source end. Constant receiver voltage is represented on the diagrams, Figs. 6-21 and 6-23 by holding a fixed origin for the analysis. Assume, for example, that V_r is to be held constant at a value of 1.0 regardless of variations in power demands by the customer. In the two diagrams, the points designated as V_r 1.0 will be used as the origin in all measurements in this section.

Assume that the load power factor is to be held constant but that the kilowatt and reactive volt-ampere loads are variable. Lines of constant power factor are shown on Figs 6-24a and 6-24b. The magnitude of the kilowatt load (in per unit) may be taken as the product of the receiver voltage

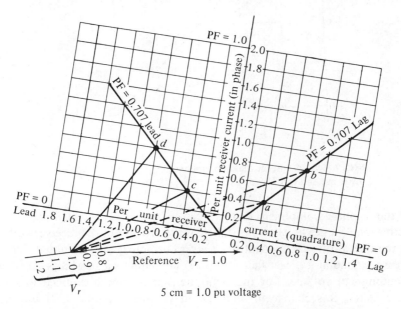

Fig. 6-24a. A voltage diagram emphasizing receiver power factor 0.707 lag and 0.707 lead.

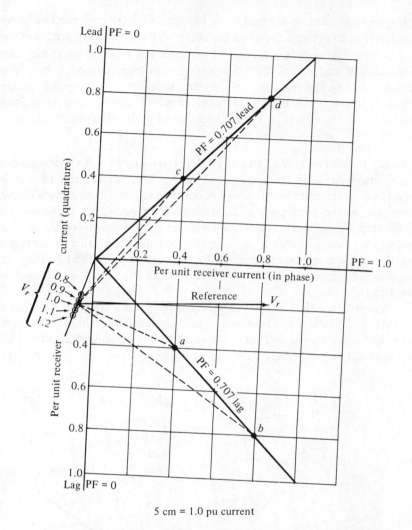

5 cm = 1.0 pu current

FIG. 6-24b. A current diagram emphasizing receiver power factor
0.707 lag and 0.707 lead.

and the in-phase component of the receiver current. The magnitude of the
reactive load (in per unit) may be taken as the product of the receiver voltage
and the quadrature component of receiver current. It may be noted that an
increase in load at a power factor of 0.707 lag requires a substantial increase
in sending-end voltage. For the same increase in power supplied at unity
power factor, less change in sending-end voltage is required. If the load
power factor is 0.707 lead, an increase in power requires almost no change
in the magnitude of the sending-end voltage. By reference to the current

diagram, Fig. 6-24b, it may be noted that the sending-end current changes
as load is changed.

For example, with a power load of 0.4 pu, power factor 0.707 lag,
point a, sending-end voltage of 1.07 pu is required. When the load is
increased to 0.8 pu, point b, the sending-end voltage must be raised to
1.34 pu. The sending-end currents for the two conditions are 0.56 and
1.07 pu.

In contrast, with a power factor of 0.707 lead, the sending voltage
for a power load of 0.4 pu, point c, is only 0.72 pu, and an increase of
load to 0.8 pu, point d, requires an increase in sending-end voltage to 0.73
pu. The sending-end currents for the two conditions are 0.86 and 1.47 pu.

With constant receiver voltage, a constant kilowatt receiver load
implies that the in-phase component of receiver current will be constant.
Assume a receiver voltage of 1.2 pu and a receiver power of 0.6 pu. The
in-phase component of receiver current will be 0.5 pu. Operation is then
along line AA on Figs. 6-25a and 6-25b, with the origins taken at V_r, 1.2.
It may be noted that as the power factor shifts from lag to lead, the magnitude of the required sending-end voltage decreases. For example, with the
above-assumed receiver voltage and kilowatt load, the required sending-
end voltage and current change markedly as the receiver quadrature power

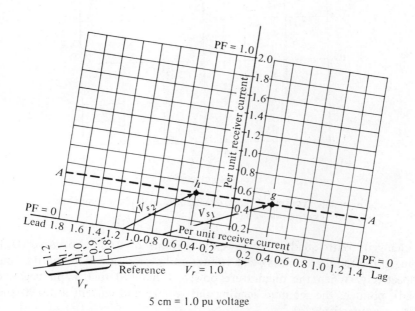

5 cm = 1.0 pu voltage

FIG. 6-25a. A voltage diagram emphasizing an in-phase component
of receiver current of value 0.5 pu.

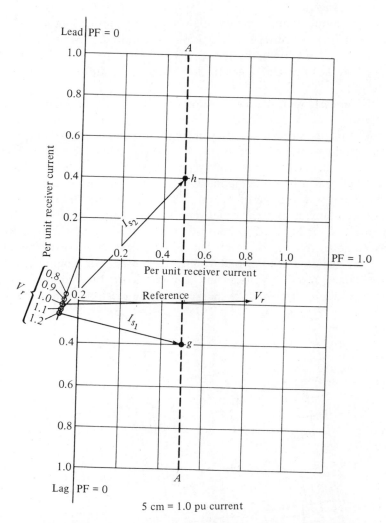

FIG. 6-25b. A current diagram emphasizing an in-phase component of receiver current of value 0.5 pu.

is changed. With quadrature power at a value of 0.48 pu positive (lag), point g, the required sending-end voltage and current are 1.26 and 0.67 pu, respectively. When the quadrature power is changed to 0.48 pu negative (lead), point h, the required sending-end voltage and current are 0.90 and 0.98 pu, respectively.

With constant receiver voltage, a constant value of receiver current implies a constant kilo-volt-ampere load, with a freedom of variation in power

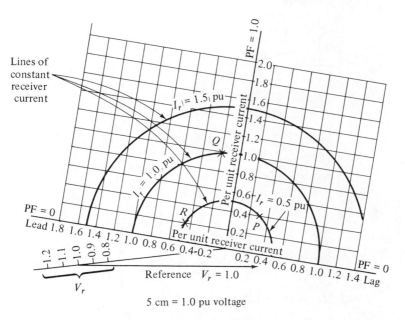

5 cm = 1.0 pu voltage

FIG. 6-26a. A voltage diagram emphasizing receiver current magnitudes of special values.

factor. The circles drawn in Figs. 6-26a and 6-26b represent this condition. Corresponding points are listed as P, Q, and R on the two diagrams.

b. Sending-End Voltage, V_s, Constant. In some instances a power system may be operated with the voltage held constant at the source end of the system. Changes in receiver voltage then result from changes in receiver power, receiver reactive power, and receiver power factor. As previously discussed, sending-end voltage is represented by a phasor drawn from a selected origin to an operating point in the current grid. If sending-end voltage is to be held constant, the length of this phasor must remain constant.

Consider the case of a variable kilowatt load at constant power factor, 0.707 lag, for example, as shown in Fig. 6-27. As load is increased, the right-hand end of the V_s phasor will move to the right while its other end will move up the AV_r line, the magnitude of V_r being determinable by the position of the lower end of the phasor V_s. When receiver voltage is 1.0 pu and receiver power is 0.1 pu (point A) the required sending-end voltage is 0.9 pu. If this sending-end voltage is held constant while receiver load (at power factor 0.707 lag) is increased until the in-phase component of current is 0.4 pu, point B, the origin of the diagram must move to point b, indicating

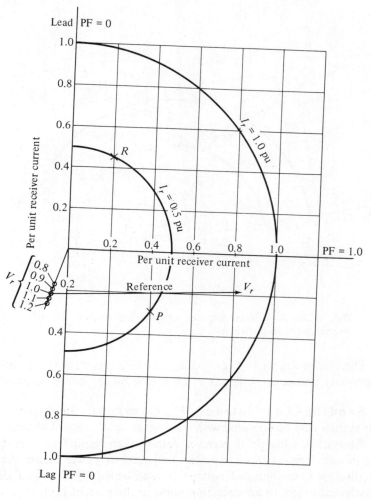

5 cm = 1.0 pu current

FIG. 6-26b. A current diagram emphasizing receiver current magnitudes of special values.

a receiver voltage of 0.75 pu. It is obvious that under this circumstance, increases in load bring about a rapid decrease in receiver voltage.

If, however, the power factor is 0.707 lead, an increase of load will cause the phasor representing V_s (shown dashed) to move in such fashion that its lower end changes position very little. With sending-end voltage held constant at 0.9 pu (as in the previous example) when the in-phase component of receiver current is 0.2 (point C), the origin of the diagram

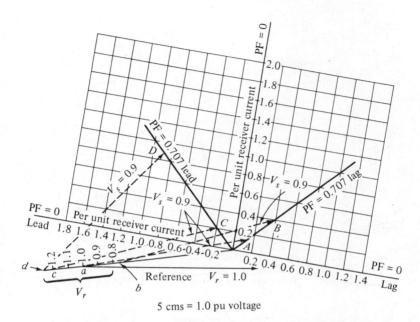

5 cms = 1.0 pu voltage

FIG. 6-27. A voltage diagram used to show changes in receiver voltage due to changes in load and power factor. Sending-end voltage V_s is held constant at 0.9 pu.

is at point c, indicating a receiver voltage of 1.18 pu. A large increase in the receiver current (to point D) shifts the origin to point d, indicating a receiver voltage of 1.25 pu. It may be concluded that if sending-end voltage is held constant on this system, increases in lagging power-factor load rapidly depress receiver voltage. Increases in leading power-factor load (at least over a limited range) cause an increase of receiver voltage. Hence for this system it may be concluded that, with a constant sending-end voltage, at lagging power factor, increases of load reduce receiver voltage much more than do increases of leading power-factor load.

With the aid of Fig. 6-28, voltage controlled by capacitor loading at the receiver end may be illustrated. Suppose that the in-phase component of receiver current remains constant at 0.7 per unit. Suppose further that initially the power factor is 0.8 lag and the receiver voltage is low, 0.8 per unit. The sending-end voltage, 1.03 pu, is then represented by the phasor PQ extending from the origin, $V_r = 0.8$, to the current indicated. Now assume capacitors are added in parallel with the load at the receiver end, thus adding negative reactive volt-amperes. With constant in-phase current, each increment of capacitor added moves the operating point to the left along the dashed line, XY. As the operating point is moved to the left, the lower end

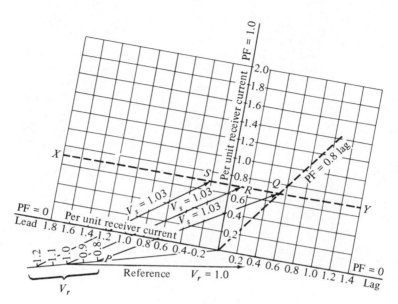

5 cms = 1.0 pu voltage

FIG. 6-28. A voltage diagram used to show changes in receiver voltage due to changes in capacitor switching at the load end. V_s is held constant at 1.03 pu.

$V_r = 1.1$ pu.

$I_r = 0.8 - j0.4$ pu.

of the phasor V_s (which is of constant length) will slide down the AV_r line defining new origins, each one corresponding to an increased value of V_r. For example, if the capacitor current is 0.8 pu, the terminal point S will be defined and a new origin corresponding to $V_r = 1.2$ pu will be required. Thus it is seen that, by switching capacitors, almost any desired value of receiver voltage may be attained. Obviously the same effect could be realized by the use of an overexcited synchronous condenser which draws a current leading its supply voltage by almost 90°.

c. Sending-End Conditions. The transmission diagrams (Figs. 6-29a and 6-29b) make possible a determination of sending-end voltage, current, and power factor once receiver-end conditions are selected.

EXAMPLE 6-6. Suppose the receiver-end voltage is 1.10 pu while receiver-end current is $0.8 - j0.4$ per unit. On Fig. 6-29a, the line MN represents the sending-end voltage in magnitude and phase position relative to V_r. In Fig.

6-29b the line PQ represents sending-end current in magnitude and phase position relative to V_r. The difference of these two angles determines the power factor at the sending end. Thus sending-end conditions are completely defined.

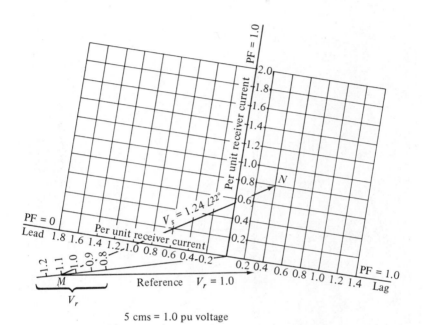

5 cms = 1.0 pu voltage

FIG. 6-29a. Determination of sending-end voltage.

In this example it may be noted that $V_s = 1.24 \underline{/22°}$ and $I_s = 0.99 \underline{/-12°}$. From these data it may be determined that at the sending end,

$$\text{power factor} = \cos(22° + 12°) = 0.83 \text{ lag}$$

$$\text{power} = V_s I_s \cos \phi_s$$
$$= 1.24 \times 0.99 \times 0.83 = 1.015 \text{ pu}$$

$$\text{reactive power} = V_s I_s \sin \phi$$
$$= 1.24 \times 0.99 \times 0.56 = 0.685 \text{ pu}$$

$$\text{system losses} = P_s - P_r = 1.015 - 0.88 = 0.135 \text{ pu}$$

d.   Other Problems. The transmission diagrams of the type just discussed will be examined in more detail in Chapter 7. With these diagrams it becomes possible to define power limits and to define criteria for stability of system operation.

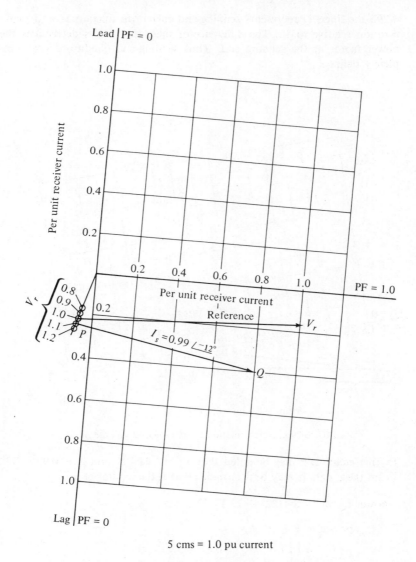

5 cms = 1.0 pu current

FIG. 6-29b. Determination of sending-end current.

6-6. SYSTEMS WITH SEVERAL SOURCES

The discussion so far has been confined to a power system that connects a single source to a single receiver. With present-day interconnection practice, many situations exist in which several sources supply one or more loads. A detailed treatment of interconnected power networks is beyond the pres-

ent objective of this discussion. However, a simple example will be presented to show that even in a multimachine case, the understanding of
simple circuit behavior is of importance.

Consider the circuit of Fig. 6-30. Here three generators, 1, 2, and 3,

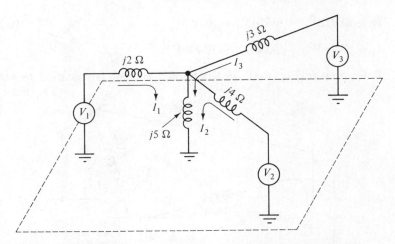

FIG. 6-30. A 3-machine interconnected system supplying a single
load.

are supplying power to a single load. It will be shown that, in effect, there
is an equivalent tie joining each pair of machines. In this illustration, let it
be assumed that the line impedances are pure inductive reactance and that
the load also is a pure inductive reactor.

The solution of this network begins by assigning the three currents
I_1, I_2, and I_3 as shown by the arrows. Then applying Kirchhoff's voltage
law around each of these defined loops, we may write

$$\bar{V}_1 = \bar{I}_1 j2 + (\bar{I}_1 + \bar{I}_2 + \bar{I}_3)j5$$

$$\bar{V}_2 = \bar{I}_2 j4 + (\bar{I}_1 + \bar{I}_2 + \bar{I}_3)j5 \qquad (6\text{-}23)$$

$$\bar{V}_3 = \bar{I}_3 j3 + (\bar{I}_1 + \bar{I}_2 + \bar{I}_3)j5$$

These equations may be rearranged and put in standard form as shown
in Eqs. (6-24):

$$\bar{V}_1 = j7\bar{I}_1 + j5\bar{I}_2 + j5\bar{I}_3$$

$$\bar{V}_2 = j5\bar{I}_1 + j9\bar{I}_2 + j5\bar{I}_3 \qquad (6\text{-}24)$$

$$\bar{V}_3 = j5\bar{I}_1 + j5\bar{I}_2 + j8\bar{I}_3$$

These three simultaneous equations may be solved for I_1, I_2, and I_3 by substitution, by determinants, or by matrix inversion with a digital computer to yield

$$\bar{I}_1 = -j0.303\bar{V}_1 + j0.0975\bar{V}_2 + j0.130\bar{V}_3$$

$$\bar{I}_2 = j0.975\bar{V}_1 - j0.201\bar{V}_2 + j0.065\bar{V}_3 \qquad (6\text{-}25)$$

$$\bar{I}_3 = j0.130\bar{V}_1 + j0.065\bar{V}_2 - j0.246\bar{V}_3$$

Refer now to Fig. 6-31. If Kirchhoff's current law is used to write

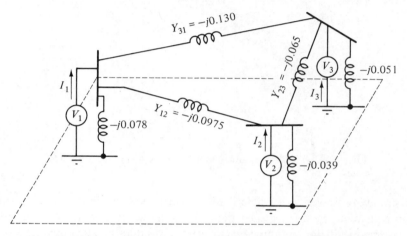

FIG. 6-31. A system equivalent to that of Fig. 6-28 as concerns current in each of the three machines. Each pair of machines is interconnected by a general power circuit.

the equations for I_1, I_2, and I_3, it will be seen that the resulting equations are exactly those shown in Eqs. (6-25). It must be concluded, therefore, that the network of Fig. 6-31 is equivalent, as far as machine terminals are concerned, to the network of Fig. 6-30.

Note that between the buses of each pair of generators there is a simple series circuit. The information we have gained regarding power flow over the general power circuit and over the generalized network helps us to understand the interchange of power between generating stations on a complicated power system.

PROBLEMS

6-1.   Refer to Fig. 6-3. Assume that
$$Z_a = 0 + j20$$
$$Z_b = 10 + j15$$

Determine Z_{eq} for this combination.

6-2.   Refer to Fig. 6-4. Assume that

$Z_a = 0 + j60$
$Z_b = 0 + j30$
$Z_c = 0 - j15$

Determine Z_{eq} for this combination.

6-3.   Refer to Fig. 6-5. Assume that

$Z_a = 15 + j0$
$Z_b = 10 + j0$
$Z_c = 60 + j0$
$Z_d = 10 + j0$
$Z_e = 10 + j0$
$Z_f = 5 + j0$
$Z_g = 15 + j0$
$Z_h = 0$

Determine Z_{eq} for this combination.

6-4.   Refer to Fig. 6-7. Assume that

$Z_1 = 0.5 + j2$
$Z_2 = 0.5 + j3$
$Z_3 = 0.5 + j3$
$Z_4 = 0.5 + j2$
$V_d = 5000 \text{ V}$
$P_a = 500 \text{ kW}, f_p = 1.0$
$P_b = 500 \text{ kW}, f_p = 1.0$
$P_c = 0$
$P_d = 0$

Solve for V_s, P_s, f_{ps}, and losses.

6-5.   Consider the two four-terminal networks 1 and 2. Determine the *ABCD* constants of the two networks in combination if

a.   1 and 2 are in tandem.
b.   2 and 1 are in tandem.
c.   1 and 2 are in parallel.

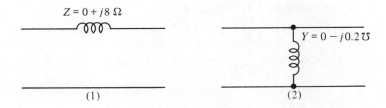

$Z = 0 + j8 \ \Omega$

(1)

$Y = 0 - j0.2 \ \mho$

(2)

6-6.   Refer to the circuit shown. Assume that V_r and I_r are of known values. Solve for V_s and I_s. From these solutions determine the $ABCD$ constants of this circuit.

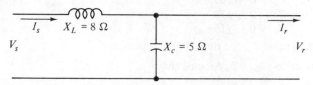

6-7.   A four-terminal network of resistors is tested with the following results:

Receiver terminals open-circuited	Receiver terminals short-circuited
$V_s = 100$ V	$V_s = 100$ V
$I_s = 15$ A	$I_s = 30$ A
$V_r = 75$ V	$V_r = 0$
$I_r = 0$	$I_r = 20$ A

Determine the $ABCD$ constants of the network.

6-8.   Consider the circuit shown. Determine the $ABCD$ constants, considering
   a. UV the source terminals, WX the receiver terminals.
   b. WX the source terminals, UV the receiver terminals.

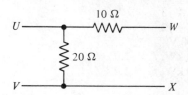

6-9.   A 105-mile 138-kV three-phase line has the following constants:
$$R = 0.168 \ \Omega/\text{mile}$$
$$X_L = 0.795 \ \Omega/\text{mile}$$
$$X_C = 187,700 \ \Omega\text{-miles}$$

   a.   Draw the π-line representation of this line.
   b.   Change the values in the representation to correspond to per unit quantities on a 50,000-kVA 138-kV base.
   c.   Construct transmission diagrams applying to this system. Use a scale of 10 cm = 1 pu voltage for the voltage diagram and 10 cm = 1 pu current for the current diagram.

6-10.   From the diagrams of Problem 6-9, make the following determinations. All are balanced three-phase conditions.
   a. $P_r = 0.90$, receiver power factor 1.0, $V_r = 1.1$. Solve for $\bar{V}_s, \bar{I}_s, F_{ps}$, and P_s.
   b. $P_{sr} = 60,000$ kVA, receiver power factor 0.8 lag, $V_r = 130$ kV. Solve for $\bar{V}_s$ and $\bar{I}_s$ in volts and amperes.
   c. $I_r = 1.0 + j0.5$ pu, $V_s = 1.1$ pu. Solve for $\bar{V}_s$ and $\bar{I}_s$.

6-11. Refer to Fig. 6-21. The system is operating with a connected load that draws a current of $0.6 - j0.4$ pu regardless of receiver voltage. The receiver voltage is 0.9 pu. It is desired to raise receiver voltage to 1.1 pu without changing the sending-end voltage. The change will be made by adding a capacitor load in parallel with connected load. What value of capacitance current will be required? How many pu ohms of capacitive reactance will this require?

6-12. Refer to Fig. 6-30. Assume that $V_1 = 10 \underline{/0°}$ V, $V_2 = 8 \underline{/0°}$ V, and $V_3 = 6 \underline{/0°}$ V. Solve for the current that flows in G_1. Make this calculation by referring to Fig. 6-31.

chapter 7

Power Limits—
Stability

7-1. STATEMENT OF THE PROBLEM

Electrical engineers in many fields of application are concerned with the maximum amount of power that can be delivered at the terminals of a circuit or a machine. The power engineer is particularly interested in this problem, for he must work under certain restrictions which do not apply to his counterpart in the communication or electronics field.

A classical problem presented in all elementary textbooks on electrical circuits is described with the aid of Fig. 7-1. In the diagram it may be seen that a voltage source V_s is connected through a resistor R to the terminals

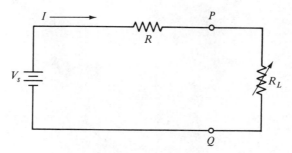

FIG. 7-1. A dc-circuit for the study of maximum power transfer.

PQ, across which is connected the variable load resistor R_L. If the voltage source is a battery, the resistor R is the sum of the internal resistance of the battery and the resistance of the connecting leads to the load terminals. The problem (as usually stated) is to determine the value of the load resistance R_L which will result in maximum power being delivered to the load. It is easily shown that the power to the load is a maximum when the load resistance R_L is equal in magnitude to the combined source and circuit resistance R,

$$R_L = R \qquad (7\text{-}1)$$

A similar problem is presented when an ac source is connected through the impedance Z_I to the load terminals PQ, Fig. 7-2. As is shown in elementary textbooks, the power delivered to the load is a maximum when

$$Z_L = \text{conjugate of } Z_I \qquad (7\text{-}2)$$

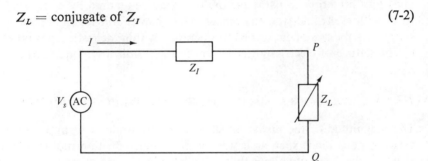

FIG. 7-2. An ac-circuit for the study of maximum power transfer.

The simple analysis of the maximum power-transfer problem as described by Figs. 7-1 and 7-2 may be adequate for the communication engineer or the electronic circuit designer. The power engineer must give attention to some additional details. In both circuits, under condition of maximum power delivery to the load, only 50 percent of the power output from the source reaches the load, a similar amount being lost in the resistor R of Fig. 7-1 or in the impedance Z_I of Fig. 7-2. In Fig. 7-1 the voltage at the load terminals is just 50 percent of that of the source. In Fig. 7-2 the voltage at the terminals of the load may be less or greater than the source voltage, depending on the nature of the impedance Z_I.

In power-system practice, loads may be in thousands of kilowatts. A system design in which, with maximum power transfer, the losses in the system were equal to the delivered load would be rejected as being uneconomical. A further complication would result because of the heat dissipated within the power-system equipment.

The power engineer is further restricted by the necessity of supplying variable loads from zero to maximum at or near a constant value of voltage. The power engineer in searching for the maximum capabilities of his system must keep in mind that the light output of lamps is greatly reduced when they are operated below rated voltage. With voltage below normal, induction motors, operating at rated torque draw abnormally high currents, which result in decreased efficiency and may result in damage from overheating. Under conditions of low voltage, starting torque is below normal, and under extreme conditions motors may stall under load.

The maximum power-transfer problem in the electric power system is further complicated by the presence of synchronous machines. If attempts are made to transfer excessive amounts of power between synchronous machines, the machines may pull out of step with each other and delivery of power to customers will be severely impaired.

In decribing power limits, it is obvious that power-system operators must give attention to many factors that may be ignored by designers who work with electronic-type equipment. This chapter will examine the power-supply capabilities of circuits when operating within the restrictions imposed by the nature of electric-power generation and utilization equipment.

7-2. ASYNCHRONOUS LOADS (LAMPS, HEATERS, INDUCTION MOTORS)

The magnitude of the power which may be transmitted from a source to asynchronous loads such as lamps, heaters, and induction motors is influenced by the range of voltage that is available by control of the source equipment, the range of voltage which may be tolerated by the load, and by the characteristics of the circuit extending from source to receiver. Restrictions may also be imposed by the current-carrying ability of the components comprising the power circuit.

The behavior characteristics of a general four-terminal network (Fig. 7-3) were discussed in detail in Chapter 6. The transmission diagram for a typical network was presented in Chapter 6 and is shown in Fig. 7-4. With the aid of this diagram, the power limits of the system to which it applies may be readily defined when source voltage, load voltage, and load power factor are specified.

As an example, suppose that receiver voltage must never fall below 0.8 pu. (This is a low value for a power application.) Assume further that

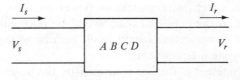

FIG. 7-3. A 4-terminal network with $ABCD$ constants.

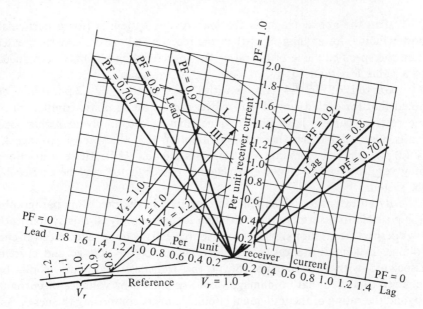

5 cm = 1.0 pu voltage

FIG. 7-4. A transmission diagram for the study of power transfer.

Curve I $V_r = 0.8$ pu $V_s = 1.0$ pu.

Curve II $V_r = 0.8$ pu $V_s = 1.2$ pu.

Curve III $V_r = 1.0$ pu $V_s = 1.0$ pu.

source voltage is held at 1.0 pu. Conforming to these restrictions the arc of a circle I may be drawn with center at $V_r = 0.8$ and with a radius of 1.0, as shown in Fig. 7-4.

Considering the load power factor to be 0.9 lag, it may be noted that for this power factor, an in-phase component of current of 0.75 pu is indicated. Hence the maximum power under these conditions is

$$P_{max} = I_{in \ phase}V_r = 0.75 \times 0.8 = 0.60 \ pu$$

Similarly, if the load power factor is assumed to be 0.707 lag, the in-phase component of the receiver current is seen to be 0.50 pu and the maximum power is

$$P_{max} = 0.50 \times 0.8 = 0.40 \ pu$$

If the load power factor is assumed to be 0.8 lead, the in-phase component of receiver current is seen to be 1.65 pu and the maximum power is

$$P_{max} = 1.65 \times 0.8 = 1.32 \ pu$$

If in the above examples the load on the system is (for a particular power factor) lower than $P_{\max}$, then the receiver voltage V_r may be greater than the specified value of 0.8 pu or the source voltage V_s may be reduced to a value less than 1.0 pu.

Curve II of Fig. 7-4 is drawn with $V_r = 0.8$ and $V_s = 1.2$ pu. It may be noted that for each assumed load power factor, the in-phase component of the receiver current is of higher value than before, implying higher load limits. Curve III is drawn on the assumption that the receiver voltage V_r is 1.0 pu, while the sending voltage V_s is 1.0 pu. Under these restrictions only very low values of power at lagging power factor may be carried by the system.

By an extension of the method described in the preceding paragraphs it is possible to determine the maximum load capability of this system with any desired restrictions placed on source-end voltage and receiver-end voltage. In some instances, a detailed analysis of the voltage and current associated with each component of the transmission circuit would be required to assure that no damage would be caused by voltages or currents beyond the rating of the individual circuit elements comprising the networks.

7-3. SYNCHRONOUS LOADS

When two or more synchronous machines are in operation on the same power system, a new limitation on maximum power transfer must be considered. This problem will be discussed in an elementary form by assuming two synchronous machines, one at the source end of our system and one at the receiver end. On examination of this system, it is found that a power-transfer limit exists, even though voltages can be held at specified values at both ends of the system. If an attempt is made to exceed this power limit, the machines will pull out of step and cease to run in synchronism. If the machine at the source end is a generator while the one at the receiver end is a synchronous motor, loss of synchronism will cause the synchronous motor to stall. If the machines at the ends of the line are both generators driven by prime movers, loss of synchronism will be followed by wild fluctuations of current and voltage (hunting) within the transmission network, accompanied by a power transfer from source to receiver which is alternately positive and negative, with an average value substantially zero. When synchronism is lost, it is invariably necessary to separate the machines by opening circuit breakers and resynchronizing them. During the period when the machines are out of step and while they are being resynchronized, customer service may be seriously impaired.

The discussion that follows refers to an analysis of the behavior of two synchronous machines operating with an intervening transmission network. The analysis will apply in principle to several different systems, such as two generators operating on the same station bus, two generators geographically separated but tied together through a transmission circuit, or a generator supplying a synchronous motor.

7-4. THE INFINITE BUS

In studying the two-machine stability problem, the analysis may be simplified considerably by assigning to one of the machines some very special and idealized characteristics. This machine, commonly referred to as an *infinite bus*, is assumed to be uninfluenced in some of its behavior regardless of the loading put upon it. It is assumed to run at constant frequency, corresponding exactly to the nominal frequency of the system. The magnitude of its voltage remains constant at any specified value and its phase position is unchanged regardless of the power and power-factor demands that may be put upon it. Because of the assumed complete stability of the voltage of the infinite bus, it is used as reference in any graphical or analytical studies of system performance.

Although the infinite bus is a fictional device used for introductory system stability analysis, there are circumstances in actual power systems which closely approximate the infinite bus concept. Present-day large interconnected power networks operate at almost constant frequency. A small generator connected to a large network would have little influence on the voltage of the network, regardless of the condition of operation of the generator. Similarly, a small synchronous motor connected to a larger interconnected system would have little influence on the voltage magnitude or phase position of the large network, even though the loading of the motor was varied over wide limits. In each of these examples the large network is a close approximation to an infinite bus.

7-5. GENERATOR CONNECTED TO INFINITE BUS

A study of the stability problem begins with an analysis of the behavior of a generator connected through a line to a infinite bus (Fig. 7-5a). In looking at the diagram of the electrical elements forming this system, it is obvious that we are dealing again with a general power circuit (Fig. 7-5b), shown in simpler form in Fig. 7-5c. For this circuit we can write the well-known equation

$$\bar{V}_s = \bar{V}_r + \bar{I}Z \qquad (7\text{-}3)$$

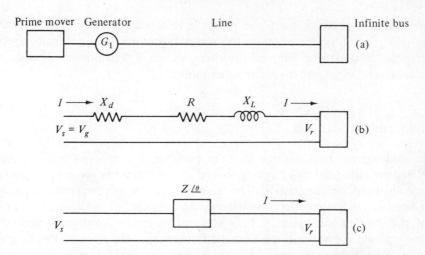

FIG. 7-5. A generator connected to an infinite bus. (a) Schematic diagram showing components. (b) Diagram showing circuit constants. (c) The system reduced to the basic power circuit.

where the voltage V_s is the internal voltage generated by the moving rotor of the generator.

a. Graphical Analysis. In Section 3-5 we developed the transmission diagram presented as Fig. 3-15. A diagram of this type (Fig. 7-6) is used here as the basis of a graphical analysis of the power-transfer characteristics of the general power circuit. This diagram is presented on a per unit basis for which $Z = 0.6 \underline{/80°}$ pu. It should be pointed out that the

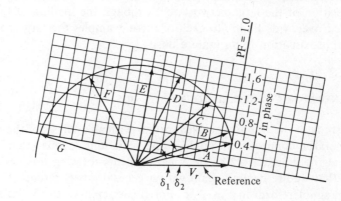

Scale 2.5 cm = 1.0 pu voltage

FIG. 7-6. A transmission diagram showing the change of $I_{\text{in phase}}$ with increase of the displacement angle δ.

(almost) horizontal lines on the grid represent constant values of the in-phase component of receiver current. Hence, for constant receiver voltage, these lines represent constant receiver power, since

$$P_r = I_{\text{in phase}} V_r$$

Suppose the system is operating under the following conditions: $V_r = 1.0$ pu, I_r (in phase) $= 0.4$ pu, power factor 1.0, $P_r = 0.4$ pu. The source-end voltage V_s is represented by the phasor A, which has magnitude 1.06 pu and lies at angle $\delta_1 = 13°$ relative to V_r.

As shown in Fig. 7-5b, the source voltage, V_s, is really the generated voltage, V_g, set up by the moving rotor in the generator. For many problems it may be assumed that this voltage generated within the machine has a magnitude that is proportional to the field current and a phase position that is dependent on the rotor position. In Fig. 7-6, V_s, represented by phasor A, is shown as leading V_r. Hence it may be concluded that the field structure of the generator is advanced by the angle δ_1 ahead of the rotor of that generator which we have called the infinite bus. Suppose that with no other change, the throttle of the prime-mover driving generator G_1 (Fig. 7-5a) is opened slowly. The increased mechanical torque produced by the prime mover at first exceeds the electrical restraining torque in the generator. Because of this unbalance of torques, the prime mover and its generator increase in speed very slightly. As the rotor (the field structure) of the generator advances in position, the generated voltage V_g will also advance to position B (Fig. 7-6) while its magnitude is unchanged.

It may be noted that in this new position (B), the in-phase component of I_r is now 0.6 pu and the power received by the infinite bus is now 0.6 pu. The increased electrical output of the generator G_1 may exactly match the increase in torque described for the prime mover. If such is the case, operation will be stable with the new displacement angle δ_2 between V_g and V_r. Since both V_g and V_r are considered to be of constant magnitude, it may be noted that the phasors A and B are radii of a circle whose center is at the origin corresponding to $V_r = 1.0$.

Further increases of the throttle opening and input power of the prime mover cause the rotor of generator G_1 to advance, and V_g takes on new positions, such as C and D. Each time the throttle is opened an additional amount, the electrical output of generator G_1 increases. The torque due to electromagnetic forces within the generator rises to meet the increased torque of the prime mover and the power delivered to the infinite bus increases. Note that in positions B, C, and D, the infinite bus is accepting power, and so has the appearance of a load on the system. This load is of leading power factor.

Suppose that by careful adjustment of the prime-mover throttle the position of the generator rotor advances until V_g is at the position E. Here the in-phase component of the current is 1.5 pu and the power received is 1.5 pu. What will be the result of a further increase of throttle opening?

An increase in the mechanical torque supplied by the prime mover again causes the rotor to speed up slightly and so advance in position relative to the rotor of the infinite bus. However, as V_g advances to a new position, F, the in-phase component of receiver current *decreases* and the power delivered to the infinite bus decreases.

As the generated voltage V_g advances in position, the output of generator G_1 decreases, the electrical torque within the machine decreases, and the amount by which the mechanical torque exceeds the electrical torque increases rapidly. This excess net torque causes a further increase in speed and a further advancement in the angle δ.

When V_g advances beyond the point G, the in-phase component of receiver current becomes negative and the infinite bus begins supplying power to the system. This further increases the acceleration of the rotor of generator G_1. The generator will increase in speed and the phasor representing V_g rotates relative to V_r with increasing speed. Synchronism between the two machines has been lost. As V_g takes on *different* positions, the currents in the circuits swing wildly and voltage measured at any point other than the terminals will rise and fall at slip frequency. The only recourse to remedy the situation is to separate the generator from the infinite bus by opening a circuit breaker, and then to resynchronize the generator to the bus.

b. Analytical Approach. The problem of the stability of two machines tied together through a general power circuit may be carried out analytically using Fig. 7-7. Note that in this diagram all complex quantities

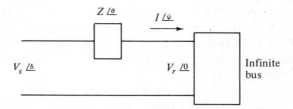

FIG. 7-7. A source voltage V_s connected through a basic power circuit to an infinite bus.

have been expressed in polar form; that is,

$$\bar{V}_r = V_r \underline{/0^\circ} \quad \text{(reference)} \tag{7-4}$$

$$\bar{V}_s = V_s \underline{/\delta} \tag{7-5}$$

$$Z = Z \underline{/\theta} \tag{7-6}$$

$$\bar{I} = I \underline{/\psi} \tag{7-7}$$

Writing Kirchhoff's voltage equation around the loop we have

$$\bar{V}_s - \bar{I}Z - \bar{V}_r = 0$$

Solving for $\bar{I}$ in this equation and expressing all terms in polar form yields

$$\bar{I} = I \underline{/\psi} = \frac{V_s \underline{/\delta} - V_r \underline{/0°}}{Z \underline{/\theta}} = \frac{V_s}{Z} \underline{/\delta - \theta} - \frac{V_r}{Z} \underline{/-\theta} \tag{7-8}$$

The power received, P_r, is the product of the *in-phase* component of the current times the scalar value of the receiver voltage:

$$I_{\text{in phase}} = I \cos \psi = \frac{V_s}{Z} \cos (\delta - \theta) - \frac{V_r}{Z} \cos \theta \tag{7-9}$$

$$P_r = I_{\text{in phase}} V_r = \frac{V_s V_r}{Z} \cos (\delta - \theta) - \frac{V_r^2}{Z} \cos \theta \tag{7-10}$$

Similarly, the reactive volt-amperes, P_q, is given as the scalar product of the *quadrature* component of the current times the receiver voltage. Hence

$$P_q = \frac{V_s V_r}{Z} \sin (\delta - \theta) + \frac{V_r^2}{Z} \sin \theta \tag{7-11}$$

If the resistance in the circuit impedance is neglected, then

$$\theta = 90° \tag{7-12}$$

Under this circumstance the expression for receiver power, Eq. (7-10), reduces to

$$P_r = \frac{V_s V_r}{Z} \cos (\delta - 90) = \frac{V_s V_r}{Z} \sin \delta \tag{7-13}$$

This sinusoidal relationship between power received and displacement angle δ is shown in Fig. 7-8. Such a plot is known as a *power–angle diagram*. This simplified version of the power–angle relationship is rather widely used in stability studies. From this curve it can be seen that maximum receiver power is attained when the displacement angle is 90°. As the angle is increased beyond 90°, receiver power drops off rapidly and becomes zero

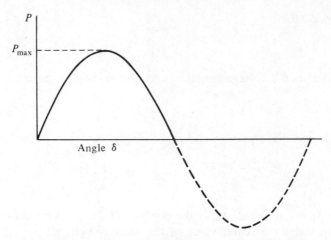

FIG. 7-8. A power-angle diagram applying to a basic power circuit in which the impedance is pure inductive reactance.

at 180°. Beyond this angle, receiver power becomes negative, which implies that the infinite bus is supplying power to the system instead of accepting power from it. It is of importance to note that if the resistance of the circuit is assumed to be zero, there are no losses in the circuit and source power is exactly equal to receiver power.

7-6. INFINITE BUS SUPPLYING A SYNCHRONOUS MOTOR

A problem similar to the one just discussed is that in which a synchronous motor is supplied from an infinite bus (Fig. 7-9a). The circuit diagram is shown in Fig. 7-9b. Here it may be noted that the source voltage is the voltage of the infinite bus, while the receiver voltage is the generated voltage within the synchronous motor. If the synchronous motor is tied directly to the infinite bus, the impedance Z is simply the impedance of the synchronous motor itself. If the motor is supplied over lines or through transformers,

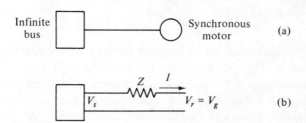

FIG. 7-9. A synchronous motor supplied from an infinite bus. (a) Schematic diagram. (b) Circuit diagram.

the impedances of these devices are included in the impedance Z shown on the diagram.

In this problem the source voltage V_s is the voltage of the infinite bus, a quantity that is assumed to remain fixed in magnitude and phase position. As the synchronous motor is loaded, its generated voltage V_g drops back in phase position.

On examination of Section 7-5b it may be noted that in the final expressions, the angle δ which appears in several equations may be interpreted as the displacement angle between the source voltage and the receiver voltage. No part of the development depended specifically on the receiver voltage remaining fixed in phase position. With this in mind, it is possible to interpret the synchronous-motor problem using the diagrams and equations developed in Section 7-5. It must be recognized that as the generated voltage of the motor drops back in phase position, it will, for purposes of analysis, continue to be used as reference. With this in mind, all diagrams and relations developed in Section 7-5 are equally applicable to the synchronous-motor problem.

7-7. STABILITY ANALYSIS FROM GENERALIZED CONSTANTS $ABCD$

In Section 7-5a, a graphical method of analysis was presented for the situation in which a single generator is connected to an infinite bus through a circuit which is reducible to the general power circuit. In some instances this simplified method of analysis is inadequate. Significant errors may be introduced if the tie between the generator and the infinite bus involves long high-voltage lines in which the shunt capacitance has considerable influence.

Regardless of the nature of the tie between the generator and the infinite bus, the circuit behavior may be analyzed by means of the generalized constants $ABCD$ (Fig. 7-10). After these constants are evaluated, a trans-

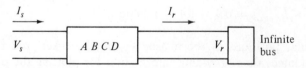

FIG. 7-10. A source V_s connected through a 4-terminal network to an infinite bus.

mission diagram similar to Fig. 6-21 (reproduced here as Fig. 7-11) may be constructed. This diagram may now be used in a fashion similar to that described in Section 7-5a. In Fig. 7-11 the arc of the circle, curve I, is drawn with V_r, the voltage of the infinite bus, assumed to be 0.9 pu, while the voltage of the source generator, V_s, is assumed to be 1.1 pu. A power–angle

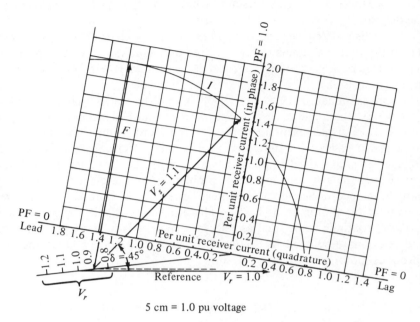

FIG. 7-11. A transmission diagram used to determine maximum power transfer.

$$V_s = 1.1 \text{ pu}$$

$$V_r = 0.9 \text{ pu}$$

diagram may be readily constructed from Fig. 7-11. One point on this diagram may be determined from the phasor V_s as drawn. It may be noted that the angle measured between V_s and the reference phasor is 45°. The in-phase component of receiver current is 1.4 pu. Receiver power is therefore

$$P_r = V_r I_{\text{in phase}} = 0.9 \times 1.4 = 1.26 \text{ pu}$$

Other angular positions of the phasor V_s may be taken and the receiver power calculated, thus providing information for plotting the power–angle diagram.

It may be noted that for this particular system and with the particular voltages specified, maximum power occurs when the V_s phasor is in position F, for here the in-phase component of receiver current is a maximum. On examination it may be noted that $P_{\text{max}} = V_r I_{\text{in phase (max)}}$ is dependent on the values assigned to V_r and V_s and on the magnitudes and angles of the constants A and B.

An analytical study of the power transfer through a generalized $ABCD$ network may be made but will not be presented here.

7-8. TRANSMISSION DIAGRAM OF A LONG TRANSMISSION LINE

Consider a power circuit in which a generator is connected to an infinite bus through a transformer and a long line as shown in Fig. 7-12. Although the transmission characteristics of this system can only be determined by

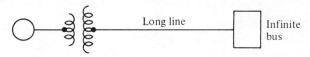

Generator   Transformer

FIG. 7-12. A generator connected through a transformer and a long line to an infinite bus.

an analysis of the total system, the transmission line itself, if long, will have a dominant influence on the behavior characteristics of the entire system. It is therefore informative to study the transmission characteristics of the line alone. A specific example will be used to illustrate the line behavior.

Consider a line having the following characteristics:

voltage	230 kV
length l	300 miles
equivalent spacing	40 ft
conductor	1,590,000 cm ACSR

Using the tables in Chapter 5 we find that

$$R = 0.0684 \ \Omega/\text{mile}$$
$$X_L = 0.806 \ \Omega/\text{mile}$$
$$X_C = 0.1908 \ \text{M}\Omega\text{-miles}$$

Considering the π-line representation of the transmission line, Eqs. (6-21) and (6-22) indicate the following:

$$Z = (R + jX_L)l = (0.0684 + j0.806)300 = 242 \ \underline{/85°}$$

$$Y_1 = Y_2 = j\frac{l}{2X_C} = j\frac{300}{2 \times 0.1908 \times 10^6} = 0.785 \times 10^{-3} \ \underline{/90°}$$

The values of A and B for the π-line representation are given by Eqs. (6-17) and (6-18) as

$$A = 1 + Y_2Z = 1 + 0.785 \times 10^{-3} \ \underline{/90°} \times 242 \ \underline{/85°}$$
$$= 1 + 0.191 \ \underline{/175°} = 0.809 \ \underline{/1°}$$

$$B = Z = 242 \ \underline{/85°}$$

With the values of A and B known, a transmission diagram may be constructed as shown in Fig. 7-13. Note that this diagram is for the single-phase equivalent circuit and is constructed in terms of volts, ohms, and amperes instead of per unit quantities. Obviously, if a base kVA had been chosen, the diagram could have been constructed in terms of per unit values. This diagram indicates that if the receiver current is 250 amperes at a power factor of 1.0 (57,500 kilowatts) the required sending-end voltage will be 200 kilovolts.

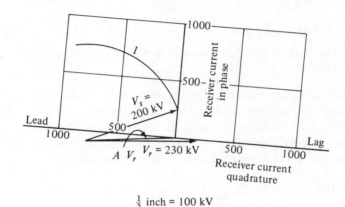

$\frac{1}{2}$ inch = 100 kV

FIG. 7-13. The transmission diagram of a 300 mile, 230 kV transmission line.

Suppose that the values of sending-end voltage V_s and receiver voltage V_r are held constant while power input is increased. The vector V_s will rotate describing the arc, I. This arc indicates a maximum possible value of the in-phase component of receiver current of 750 amperes (172,000 kilowatts).

To make better use of his equipment, the transmission engineer looks for methods of increasing the power limits of his system. This search has led to many improvements in transmission-circuit design.

7-9. METHODS OF INCREASING TRANSMISSION-SYSTEM CAPABILITIES

Various methods are available for increasing the transmission capabilities of circuits. Some of these methods are illustrated with reference to the transmission line described in Section 7-8, even though the line is only a part of the necessary system.

a. Multiple Circuits. Referring to the line described in Section 7-8, it is obvious that building a second circuit would bring about a doubling of the transmission capabilities, for each line would be capable of carrying the maximum power, as previously described. Multiple circuits are in common use. They have the additional advantage that service may be maintained over one line, with reduced capability, while the other line is out of service for maintenance or repair.

b. Raising Line Voltage. Referring to Fig. 7-13, it may be noted that if the receiver voltage V_r is increased, the phasor AV_r increases proportionately. Assuming again an initial loading of 57,500 kilowatts at unity power factor, receiver current decreases but V_s increases. This increased value of V_s provides an increased maximum value of the in-phase component of receiver current. Combining this with increased receiver voltage results in increased maximum power transfer.

c. Use of Shunt Reactors. Shunt inductive reactors of proper size may be installed at each end of the line (Fig. 7-14), thereby effectively balancing out the line capacitance and reducing the value of Y in Eq. (6-22) to zero. Under this circumstance the value of A becomes 1.0 and the transmission diagram changes to that shown in Fig. 7-15.

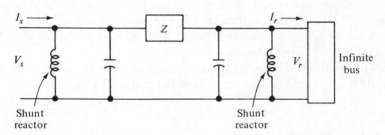

FIG. 7-14. The π-line representation of the 230 kV transmission line with shunt reactors at the terminals.

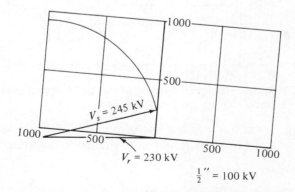

FIG. 7-15. The transmission diagram of the 230 kV transmission line with shunt reactors.

Again assuming a receiver voltage V_r of 230 kilovolts and a receiver current of 250 amperes, unity power factor (57,500 kilowatts), it may be noted that V_s is 245 kilovolts, a value substantially higher than that required as indicated in Fig. 7-13. Operating with this value of sending-end voltage, power transfer may be increased until the in-phase component of receiver current is 850 amperes, corresponding to a maximum power limit of 196,000 kilowatts. The effect of the shunt reactors is even more favorable if the analysis is made considering the impedance of the generator and transformer at the source end.

When the system is equipped with reactors as described above, the source generator must operate at a higher voltage, which means that it has a stronger field. A high value of field strength tends to provide stable operation.

It may be noted that the long line with shunt reactor compensation has the behavior characteristics of the general power circuit.

d. Reducing Line Reactance By Means of Series Capacitors. It may be noted that in the construction of transmission diagrams for either the short-line representation or the π-line representation of a transmission line, the spacing between the lines of the grid is determined by the series impedance of the line. Any means of reducing this impedance reduces the spacing between grid lines, from which it follows that power limits are increased.

One method of reducing the series impedance is installation of series capacitors in the line, usually located at the midpoint (Fig. 7-16).

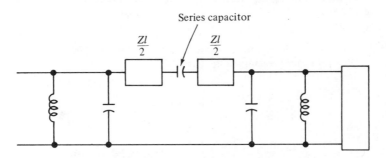

FIG. 7-16. The π-line representation of the 230 kV transmission line with shunt reactors and a series capacitor.

Referring to the transmission line of Section 7-8, the installation of a series capacitor of 120 ohms results in a total series impedance of

$$Z_t = (R + jX_L)l - jX_C$$
$$= (0.0684 + j0.806)300 - j120$$
$$= 20.5 + j122 = 124 \; \underline{/80.4^\circ}$$

With reactor compensation also (as described in Section 7-9c) the transmission diagram appears as shown in Fig. 7-17. Note that for the same initial load conditions (57,500 kilowatts at 230 kilovolts) the maximum power transfer is 390,000 kilowatts.

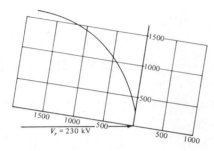

FIG. 7-17. The transmission diagram of the 230 kV transmission line with shunt reactors and a series capacitor.

e. Reducing Line Reactance By Means of Bundle Conductors. Although the use of bundle conductors on transmission lines is primarily to reduce corona loss and radio interference, it also has the advantage of reducing line reactance and so increasing the power limit of the line. Suppose a conductor of a given physical radius (and with a particular value of geometric mean radius GMR_1) is replaced by four cables of equal total conductance spaced apart as described in Section 5-2e. The new value of the geometric mean radius GMR_{eq} as given by Eq. (5-9) is greater than GMR_1. Hence the inductive reactance as given by Eq. (5-4) is reduced slightly, with a corresponding increase in the maximum power-transfer capability of the line.

f. Use of Voltage Regulators. Automatic voltage regulators may substantially increase the steady-state stability limits of a system, as will be illustrated with Fig. 7-18. In (a) is shown the circuit to be considered. A generator connects to a low-voltage bus, which in turn connects through a transformer bank to a high-voltage bus and a long transmission line.

In discussions so far we have assumed that the source voltage V_g (Fig. 7-18b) is held constant while we observe power changes as displacement angle is advanced. It is a good approximation to assume that voltage V_g is proportional to field current and remains constant if field current is held constant. If power–angle studies are to be made under this circumstance, it is proper to consider the entire series impedance between V_g and the infinite bus.

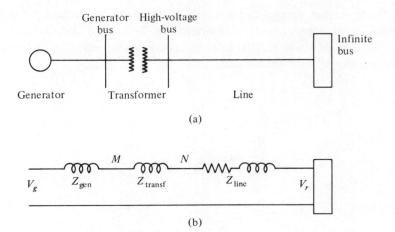

FIG. 7-18. A simple power system showing all buses. (a) Schematic.
(b) Circuit diagram.

In power-system practice it is quite common to readjust field current and so change V_g as loading on the machine is increased. This change in field current may be accomplished by an automatic voltage regulator. If the automatic voltage regulator is set to hold a constant value of voltage on the generator bus (point M, Fig. 7-18b), power limits are determined by a consideration of the series impedance from the point M to the infinite bus. Obviously because of the lower value of the total series impedance Z, this mode of operation will give higher power limits than would be obtained by holding V_g (field current) constant.

A still more favorable situation is observed if the voltage regulator controlling the field current of the generator is set to hold at a constant value the voltage on the high-voltage bus, N. Now the only impedance that needs to be considered in the problem is that which lies between point N and the infinite bus.

It is the operating practice of some companies to set their voltage regulators to hold constant a voltage which may be regarded as at a point somewhere between M and N. Local considerations determine the exact point to be held constant by action of the voltage regulator.

If, in the example just cited, the power output of the generator is to be increased while an automatic voltage regulator holds generator bus voltage at a constant value, it is necessary for field current to be increased with each increment of power output. In some instances it is impossible to realize the full power limit predicted by graphical analysis or by the equations because of field-circuit limitations. In every machine the maximum value of V_g is limited by the maximum possible output voltage of the excitor supplying the field, by the resistance of the field circuit, and by the saturation of the magnetic circuit. All these factors must be evaluated when one attempts to determine the steady-state power limits of a particular machine.

7-10. TRANSIENT STABILITY

The discussion in this chapter up to the present time has been limited to the steady-state performance of electric power circuits. The study of this type of behavior is of importance in predicting the response of a system on which load changes are made very gradually. This type of analysis is also of importance as an introduction to the behavior of electric power systems when subjected to sudden load changes or to sudden changes in circuit configuration. Some of the elementary problems of transient analysis will be introduced in this section.

Consider the system shown in Fig. 7-19, in which the problem is

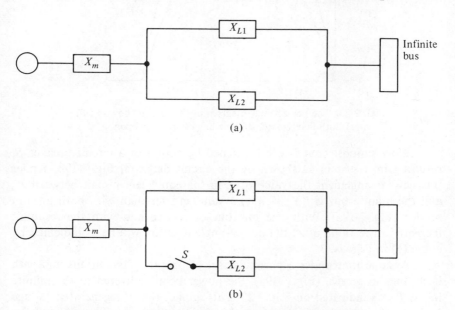

(a)

(b)

FIG. 7-19. A generator connecting through two parallel lines to an infinite bus. (a) Both lines closed. (b) One line closed, one open.

simplified by considering reactances only. This diagram represents a generator connecting to a bus and then through a parallel combination of two lines to an infinite bus. X_m represents the reactance of the machine and X_{L1} and X_{L2} the reactances of the two lines.

By suitable circuit reduction the system of Fig. 7-19a may be reduced to the general power circuit in which a single impedance Z appears between the generated voltage V_g and the infinite bus (Fig. 7-7). Assuming that V_s, V_r, and Z are known, reference to Eq. (7-13) permits the plotting of a power–angle diagram as shown by (a) in Fig. 7-20.

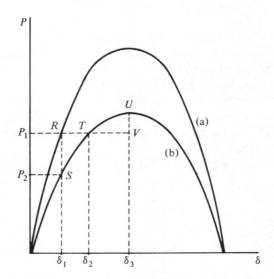

FIG. 7-20. The power-angle diagram of the system shown in Fig. 7-19. (a) Both lines closed. (b) One line closed, one open.

Now suppose that line 2 is opened by action of a circuit breaker S, resulting in a system as shown by the circuit diagram, Fig. 7-19b. Under this new arrangement the value of Z in the equivalent circuit between V_g and the infinite bus is $j(X_m + X_{L1})$ a value greater than with both lines in service (Fig. 7-19a). With one line out of service and with the resultant increase in the value of Z, the power–angle diagram will be as indicated by curve (b) in Fig. 7-20.

Assume that under a particular set of operating conditions and with both lines in service (Fig. 7-19a), the power being delivered to the infinite bus is P_1, as indicated on Fig. 7-20. This implies that the generated voltage V_s leads the voltage of the infinite bus by the angle δ_1. The rotor of the machine then is at a particular location with respect to the rotor of the infinite bus.

Next suppose that the switch S suddenly opens, giving us the circuit indicated by Fig. 7-19b. Since the angular displacement of the rotor cannot change instantly, immediately after the opening of the breaker the rotor displacement is still at the angle δ_1, which implies that the power delivered to the load and the power output of the generator has dropped to the value P_2. This sudden change in the output of the generator could not be immediately recognized by the governor on the prime mover, with the result that shaft torque will remain unchanged following the switching operation. Now the power input to the generator from the shaft is P_1 while the power output into the electric circuit is P_2, the difference between the two being available for accelerating the rotor. With a slight increase in speed, the rotor angle will increase, eventually reaching the value at which shaft power and electrical power will again balance. However, during the time the shift has been made from δ_1 to δ_2, the rotor has been accelerating. When the angle δ_2 is reached, the rotor is running slightly above synchronous speed. It will therefore continue to advance in the direction of δ_3. However, in the region between δ_2 and δ_3 the electrical output exceeds the shaft input, with the result that the rotor slows down.

It can be shown for this particular case (although it will not be done here) that the rotor will advance until the area TUV is equal to the area RST. Oscillation will be observed around the new point of stability, δ_2, with eventual stable operation at this rotor displacement.

The study of transient stability problems involves an analysis of electric-circuit behavior combined with considerations of transients in rotating mechanical systems. This subject is treated in detail in more advanced courses and textbooks.

7-11. Systems with Many Machines

Present-day power systems have extensive interconnections which result in many synchronous machines, generators, and motors, being electrically connected through complicated networks.

The analysis of the two-machine problem as presented above is an oversimplification of the case, inasmuch as one machine was presumed to be of infinite size. If two machines of finite size are considered, any change in either of the machines affects the other with complicating results. More advanced methods than have been presented here must be used even in this simple case.

Large electric power networks bring about the interconnection of many machines in different plants at widely separated points. The steady-state and transient analysis of such systems, involving many lines and transformers, is a highly specialized subject. Methods of study using digital computers must be continually revised as system growth adds to the complexity of the problem.

PROBLEMS

7-1.   Determine the maximum power that may be delivered by a 12-V storage
battery which has an internal resistance of 0.05 Ω. Under maximum power
conditions, what will be the power loss within the battery and what will be
the terminal voltage?

7-2.   Determine the maximum power that may be delivered by a 10-kVA transformer whose primary voltage is held constant at 1.1 pu volts. The impedance of the transformer is 2 percent resistance and 4 percent reactance.
Under maximum power conditions, what will be the power loss within the
transformer, the terminal voltage, and the current?

7-3.   Refer to the transmission diagram presented in Fig. 6-21. With $V_r = 1.0$ pu,
draw curves showing the magnitude of V_s for receiver power P_r ranging from
0 to 1.2 pu, the power factor assumed to be

   a.   unity.
   b.   0.707 lag.
   c.   0.707 lead.

7-4.   For the system of Problem 7-3, assume that $V_s = 1.1$. Draw curves of V_r
for receiver power ranging from 0 to 1.2 pu, the power factor assumed
to be

   a.   unity.
   b.   0.8 lag.
   c.   0.8 lead.

7-5.   The system of Problem 7-3 is to carry a unity-power-factor customer load
which varies from 0 to 1.6 pu. Sending-end voltage is held constant at
1.1 pu. Receiving-end voltage is held constant at 1.0 pu by adjusting a
reactive load which is in parallel with the customer load. Determine the
reactive volt-amperes (lag or lead) necessary if the customer load is

   a.   0.
   b.   0.5 pu.
   c.   1.0 pu.
   d.   1.3 pu.

7-6.   A generator with an internal impedance of $(0.05 + j.80)$ pu is connected
to an infinite bus, whose voltage is 1.0 pu. Construct a transmission diagram
and draw a power angle curve assuming that V_g is 1.1 pu. What is the
maximum power received and what will be the circuit loss at this power?

7-7.   Refer to Problem 7-6. Assume that the generator is connected to the infinite
bus through a line whose impedance is $Z_L = 0.2 + j0.5$. What is the maximum power that may be transferred to the infinite bus?

7-8.   Refer to the transmission diagram Fig. 6-21 and consider it applying to a
generator connected to an infinite bus. Assume that the infinite bus voltage
is 1.1 pu. Calculate the rotor-displacement angle when the received power

is 0.8 pu, assuming that the generator voltage is

a.   0.95 pu.

b.   1.10 pu.

7-9.   Compare the maximum power transfer for the transmission line described in Section 7-8 with and without shunt reactors. Assume that in each case $V_s = 245$ kV and $V_r = 230$ kV.

7-10. Refer to the capacitor-compensated line in Section 7-9d. Suppose that by mistake the capacitive reactance is made 300 Ω. Draw a transmission diagram for this condition, and from it analyze system performance.

7-11. Refer to the system of Fig. 7-19 and the power angle diagram Fig. 7-20. Assume that after switch S is opened, all transients have disappeared and the system is operating stably at the angle δ_2. Switch S is now closed. Analyze the system response.

chapter 8

Faults on
Power Systems

Each year new designs of power equipment bring about increased reliability of operation. Nevertheless, equipment failures and interference by outside sources occasionally result in faults on electric power systems. On the occurrence of a fault, current and voltage conditions become abnormal, the delivery of power from the generating stations to the loads may be unsatisfactory over a considerable area, and if the faulted equipment is not promptly disconnected from the remainder of the system, damage may result to other pieces of operating equipment.

8-1. DESCRIPTION OF A FAULT

A *fault* is the unintentional or intentional connecting together of two or more conductors which ordinarily operate with a difference of potential between them. The connection between the conductors may be by physical metallic contact or it may be through an arc. At the fault, the voltage between the two parts is reduced to zero in the case of metal-to-metal contacts, or to a very low value in case the connection is through an arc. Currents of abnormally high magnitude flow through the network to the point of fault. These short-circuit currents will usually be much greater than the designed thermal ability of the conductors in the lines or machines feeding the fault. The resultant rise in temperature may cause damage by the annealing of conductors and by the charring of insulation. In the period during which the fault is permitted to exist, the voltage on the system in the near vicinity of the fault will be so low that utilization equipment will be inoperative. It is apparent that the power-system designer must anticipate points at which faults may occur, be able to calculate conditions that exist during a fault, and provide equipment properly adjusted to open the switches necessary to disconnect the faulted equipment from the remainder of the system. Ordinarily it is desirable that no other switches on the system are opened, as such behavior would result in unnecessary modification of the system circuits.

A distinction must be made between a *fault* and an *overload*. An overload implies only that loads greater than the designed values have been imposed on the system. Under such a circumstance the voltage at the overload point may be low, but not zero. This undervoltage condition may extend for some distance beyond the overload point into the remainder of the system. The currents in the overloaded equipment are high and may exceed the thermal design limits. Nevertheless, such currents are substantially lower than in the case of a fault. Service frequently may be maintained, but at below-standard voltage.

Overloads are rather common occurrences in homes. For example, a housewife might plug five waffle irons into the kitchen circuit during a neighborhood party. Such an overload, if permitted to continue, would cause heating of the wires from the power center and might eventually start a fire. To prevent such trouble, residential circuits are protected by fuses or circuit breakers which open quickly when currents above specified values persist. Distribution transformers are sometimes overloaded as customers install more and more appliances. The continuous monitoring of distribution circuits is necessary to be certain that transformer sizes are increased as load grows.

Faults of many types and causes may appear on electric power systems. Many of us in our homes have seen frayed lamp cords which permitted the two conductors of the cord to come in contact with each other. When this occurs, there is a resulting flash, and if breaker or fuse equipment functions properly, the circuit is opened.

Overhead lines, for the most part, are constructed of bare conductors. These are sometimes accidentally brought together by action of wind, sleet, trees, cranes, airplanes, or damage to supporting structures. Overvoltages due to lightning or switching may cause flashover of supporting insulators, thereby establishing an arc from the conductor to the support or from conductor to conductor. Contamination on insulators sometimes results in flashover even during normal voltage conditions.

The conductors of underground cables are separated from each other and from ground by solid insulation, which may be oil-impregnated paper or a plastic such as polyethylene. These materials undergo some deterioration with age, particularly if overloads on the cables have resulted in their operation at elevated temperatures. Any small void present in the body of the insulating material will result in ionization of the gas contained therein, the products of which react unfavorably with the insulation. Deterioration of the insulation may result in failure of the material to retain its insulating properties, and short circuits will develop between the cable conductors. The possibility of cable failure is increased if lightning or switching produces transient voltages of abnormally high values between the conductors.

Transformer failures may be the result of insulation deterioration combined with overvoltages due to lightning or switching transients. Short circuits due to insulation failure between adjacent turns of the same winding may result from suddenly applied overvoltages. Major insulation may fail, permitting arcs to be established between primary and secondary windings or between a winding and grounded metal parts such as the core or tank.

Generators may fail due to breakdown of the insulation between adjacent turns in the same slot, resulting in a short circuit in a single turn of the generator. Insulation breakdown may also occur between one of the windings and the grounded steel structure in which the coils are embedded. Breakdown between different windings lying in the same slot results in short-circuiting extensive sections of the machine.

Fault may be classified as permanent or temporary. *Permanent faults* are those in which insulation failure or structure failure produces damage that makes operation of the equipment impossible and requires repairs to be made. *Temporary faults* are those which may be removed by deenergizing the equipment for a short period of time; short circuits on overhead lines frequently are of this nature. High winds may cause two or more conductors to swing together momentarily. During the short period of contact, an arc is formed which may continue as long as the line remains energized. However, if automatic equipment can be brought into operation to deenergize the line quickly, little physical damage may result and the line may be restored to service as soon as the arc is extinguished. Arcs across insulators due to overvoltages from lightning or switching transients usually can be cleared by automatic circuit-breaker operation before significant structure damage occurs. Because of this characteristic of faults on lines, many companies operate following a procedure known as high-speed reclosing. On the occurrence of a fault, the line is promptly deenergized by opening the circuit breakers at each end of the line. The breakers remain open long enough for the arc to clear, and then reclose automatically. In many instances service is restored in a fraction of a second. Of course, if structure damage has occurred and the fault persists, it is necessary for the breakers to reopen and lock open.

For the purpose of this discussion, one specialized type of fault is given primary consideration, the balanced three-phase fault. This type of fault exists when all three conductors of a three-phase line (or a machine) are brought together simultaneously into a short-circuit condition. Faults of this type are not particularly common, but their study enables us to comprehend a whole group of problems that short-circuit conditions present to the power-system operator. Other types of faults, which are much more difficult to analyze, are discussed briefly in Section 8-6.

Balanced three-phase faults, like balanced three-phase loads, may be handled on a line-to-neutral basis or on an equivalent single-phase basis. Problems may be solved either in terms of volts, amperes, and ohms, or in terms of per unit quantities. The handling of faults on single-phase lines is of course identical to the method of handling three-phase faults on an equivalent single-phase basis.

8-2. NEED FOR CALCULATING FAULT CONDITIONS

Faults on power systems are unpredictable both as to location and time of occurrence. When a system study is made to determine the behavior during a fault, it is necessary to assume the location of the fault, the configuration of lines, transformers, and generators that exists previous to the fault, and in some cases the loading of the system. For a system of some complexity, the possibilities of assignment of initial conditions combined with a choice of fault location may result in a very large number of required solutions. These studies provide the engineer with information by which he can design to assure the prompt disconnection of faulted equipment with a minimum of damage and a minimum of disturbance to the operation of the remaining system.

Circuit breakers and fuses are used to disconnect faulted equipment. Since the current to the faulted equipment is abnormally high, these devices are asked to interrupt currents far greater in magnitude than those observed during normal operation. The circuit-interrupting process consists of parting a pair of contacts between which an electric arc is drawn, then processes must be brought into play to extinguish the arc. The greater the arc current and the higher the voltage of the circuit, the more difficult is the problem of arc extinction. Breakers and fuses are rated in terms of the normal circuit voltage, the continuous currents they may carry, and the short-circuit currents they may interrupt. Under short-circuit conditions a circuit breaker may open successfully, provided the current to be interrupted is within the design value. However, when asked to interrupt a current well above the design value, the arc between the parting contacts may not extinguish and the breaker may be destroyed by gas pressure built up within it (Fig. 8-1). Ordinary

FIG. 8-1. A circuit breaker damaged by short-circuit currents which exceeded its interrupting rating.

plug fuses, such as those used in entrance boxes in residences, will interrupt currents far above their continuous rating. On typical residential circuits, such fuses, when short circuits occur, will blow without breaking their transparent windows. However, if misapplied on industrial circuits where short-circuit currents may be thousands of amperes, the blowing of a fuse may produce the sound and violence of a shotgun blast.

As systems grow and new generators, lines, and transformers are added, short-circuit current values increase. It may become necessary to replace breakers and fuses with ones of higher current-interrupting capacity to avoid failures.

In order that circuit breakers operate promptly to disconnect faulted equipment, it is necessary to have a system that may recognize the presence of a fault, determine which piece of equipment is faulted, and supply energy to the mechanisms which will trip the appropriate breakers. In some of the simplest systems using low-voltage breakers, the power current itself flows through the trip coil of the breaker. If the current exceeds a certain specified value, the breaker is tripped and the circuit opened. On complicated networks, a system of on-line computers (known to the power man as relays) receive a continuous indication of the current and voltage on all lines, transformers, and generators connected to the many station buses. On the occurrence of a fault, the fault condition is recognized by the abnormal voltage and current conditions then in existence. The relay contacts close to supply tripping current to those breakers necessary to clear the fault. If these relays are to be adjusted for proper functioning, it is necessary for the power engineer to know in advance the values and phase positions of the currents and voltages resulting from fault conditions. The subject of relays is treated in more detail in Chapter 11.

A force exists between a current-carrying conductor and every current-carrying conductor near it. These forces must be given consideration in the design of station buses, transformer windings, and machine windings. For two parallel conductors the force per unit length is directly proportional to the product of the current in each of the two conductors and inversely proportional to their separation. Equipment must be designed to withstand the forces resulting from the extremely high currents that exist during faults.

Overhead conductors, underground cables, transformer windings, and generator coils may be damaged by excessive temperature, either through the annealing of the conductor metal or through damage to insulation. In some cases the current during a fault, even though promptly cleared, may be so great that overheating of a conductor may occur.

8-3. FAULT CONDITIONS FROM CIRCUIT PARAMETERS

The circuit conditions during a fault may be calculated directly from known circuit parameters. On simple circuits, the problem is straightforward and easily accomplished. On complicated networks, digital computers are used to solve the required mathematical relations. A few examples will demonstrate the general method of attack.

a. Simple Series Circuit. A circuit representative of a distribution line is shown in Fig. 8-2a. Assume that the sending-end voltage is held constant at 4800 volts and the line impedances are as indicated. Load points are at A, B, and C.

In many instances the short-circuit currents which flow during faults are much greater than normal load currents, permitting load currents to be neglected (Fig. 8-2b).* Faults may be simulated at a station by closing switch P, Q, or R. Table 8-1 shows the system conditions existing for a fault at each of the three points, A, B, and C. Obviously faults could occur between stations. The resulting conditions at the stations may be determined once the location of the fault point is assumed.

---

*An important exception is synchronous motor load. These motors must be considered as generators during faults.

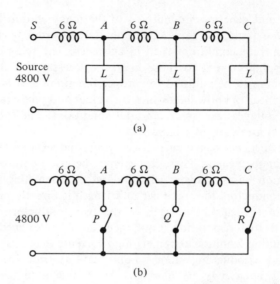

FIG. 8-2. A distribution circuit with 3 load points. (a) Normal circuit. (b) Loads are ignored. Short circuits may be produced by closing switches P, Q, or R.

Table 8-1. *Fault Condition on Fig. 8-2*

	Fault at A	Fault at B	Fault at C
CURRENT (amperes)			
S to A	800	400	266
A to B	0	400	266
B to C	0	0	266
VOLTAGE (volts)			
At A	0	2400	3200
At B	0	0	1600
At C	0	0	0

b. Faults on Networks. Figure 8-3a shows a simple network consisting of a transformer bank supplying a load over two parallel lines, on one of which a fault is assumed to exist. Figure 8-3b shows the assumed values of circuit constants in per unit values. The load is again neglected. This circuit may be reduced by a series of steps, shown in Fig. 8-3c and d, to a simple circuit in which the fault current may be determined. Once the magnitude of fault current is known, its division between the two line sections may be determined. Finally, the voltage at each bus and the current in each line may be evaluated as shown in Fig. 8-3e.

So far the discussion has centered on the calculation of currents and

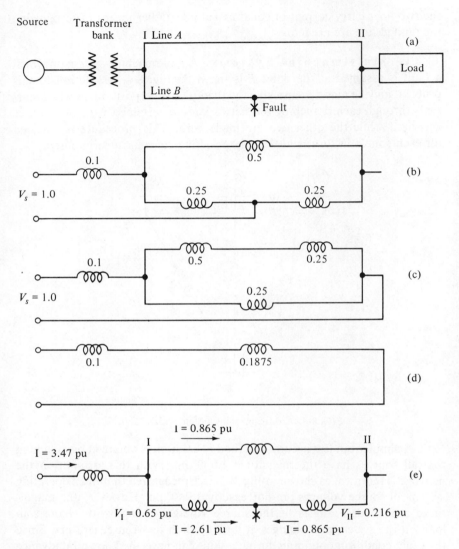

FIG. 8-3. A typical small power system with fault. (a) One line diagram. (b) Single-phase equivalent circuit under fault conditions. (c) and (d) Steps in system reduction for fault calculations. (e) Calculated currents in each line and voltages at each bus.

voltages in the presence of a fault at a known location. When an actual fault occurs, usually its location is unknown and the piece of faulted equipment must be identified by the currents and voltages observed at the stations during the fault. From Fig. 8-3e it is obvious that the faulted line could be identified at bus I simply on the basis of current magnitudes. At bus II, the two lines carry current of equal magnitude. Other means of recognizing the faulted line are needed.

c. Multimachine Systems. A multimachine system on which a fault is assumed at the point F is shown in Fig. 8-4. A solution to this problem may be made by applying the theory of superposition, in which currents through each branch of the network are determined with one source energized while the other two are made zero. This procedure is repeated for each source in turn, and then the resulting currents superimposed.

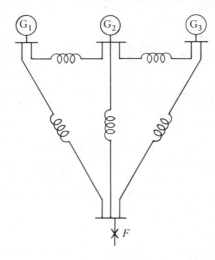

FIG. 8-4. A three-machine system with a fault.

A simpler but less accurate method of solution is obtained by assuming that all sources have the magnitude 1.0 $\underline{/0°}$ pu. With this assumption the network is redrawn, as shown in Fig. 8-5. It is recognized that in an operating system all source voltages are not exactly 1.0 $\underline{/0°}$ pu. However, the magnitudes and relative angles of the source voltages change with changes in load on the system and changes in load division between generators. Since the exact condition obtaining during a fault can never be known in advance (and will change from one fault to the next) the assumption of unit voltage on each machine is well justified for most short-circuit calculation.

Another simplifying assumption stems from the fact that in most power-system components the ratio between X_L and R is very large. This fact permits the resistance to be neglected without a significant loss in the accuracy of results, but with considerable simplification in the computations. However, this approximation must be used with caution in calculations pertaining to systems that have considerable circuit mileage in cables. The spacing between conductors in insulated cables is small, resulting in a low inductive reactance per mile. Thus the X/R ratio will be much smaller than that found in transformers, generators, and overhead lines, and the omission of the cable resistance may result in errors of significance.

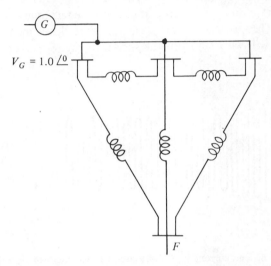

$V_G = 1.0 \angle 0$

FIG. 8-5. The three-machine system of Fig. 8-4 simplified by assuming a voltage of 1.0 $\underline{/0°}$ at each source.

8-4. SYNCHRONOUS MACHINE IMPEDANCE

In Section 2-8a the equivalent circuit model (or Thévenin equivalent) of an ac generator was shown (Fig. 2-48) as a generated voltage connected to the machine terminals through a resistance and an inductive reactance. This is, indeed, a gross oversimplification of a rather complicated electrical machine. However, this model serves for many purposes, particularly if the constants in the equivalent circuit are chosen to fit the particular problem at hand. Table A-7 was presented with but little discussion of the data. The application of the various values of inductive reactance listed in this table will now be presented. The value to be selected depends upon whether conditions of steady state or rapid change are being considered.

a. Problems of Steady State. Steady-state conditions are assumed in those problems in which changes of load are made very gradually, such as over a period of several minutes or more. Studies pertaining to voltage regulation and to steady-state stability limits fall in this class. For problems of this type it is proper to use the inductive reactance given in Table A-7 under the heading X_d. This refers to the direct axis reactance or, as some writers prefer to call it, the synchronous react-ance, X_s.

b. Calculations Involving Fault Conditions. Sup-pose that a short circuit is suddenly applied to the terminals of a three-phase ac generator. The power-frequency current that flows in any one of the three windings will vary with time as shown in Fig. 8-6. Note that the cur-

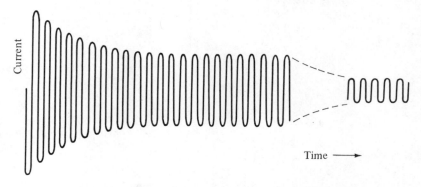

FIG. 8-6. The current in a short-circuited ac-generator decreases with time.

rent is initially of high magnitude, falls in value rapidly in the first 5 to 15 cycles, and eventually comes to a steady value 2 or 3 seconds after the start of the fault. This actual performance is not that which would be expected from a consideration of the equivalent circuit of the alternator. However, the equivalent circuit will predict expected current values at any desired point in time if the proper value of alternator impedance is selected.

It is frequently desirable to determine the 60-hertz component of current which exists *immediately* after the initiation of the fault. This current influ-ences the relays that provide system protection, it determines the electro-magnetic forces set up, and, if breakers open with extremely high speed, is the current which must be interrupted. In the solution of problems of this sort it is necessary to use the inductive reactance X_d'', which is designated as the *subtransient reactance*. It may be noted that for all machines, X_d'' is *much smaller* than X_d.

In certain power-system problems, attention is given to the magnitude of the current 10 or 12 cycles after the initiation of the fault. Intentional time delay in the operation of relays and circuit breakers may result in this current being of importance in determining the current interrupting ratings of breakers. In the case of a circuit protected by a fuse, a short time period is necessary to raise the temperature of the fuse to the melting point. The current which the fuse must interrupt is that which exists after the fuse has melted and an arc has been established. Hence it is necessary to calculate the value of the current in a circuit after the early rapid decrease in current value. For problems of this type, the *transient reactance* X_d' should be used in calculations. As would be expected, the value of X_d' is intermediate between the values of X_d and X_d''.

When a short circuit is permitted to continue for an extended period of time, transients disappear and the current magnitude reaches a steady state. For calculations of current magnitude under this condition, the generator is assigned the impedance X_d, as indicated in Section 8-4a.

The phenomena that occur within a synchronous machine which make necessary the use of these three different values of reactance involve transients in the field and in the rotor of the ac machine. An analysis of this behavior is beyond the scope of the present discussion.

8-5. DC OFFSET

In Section 8-4 we described the behavior of an ac generator with regard to power-frequency currents and their variation with time following the occurrence of a short circuit. Another effect, known as *dc offset* when applied to ac generators, is common to all inductive circuits subjected to sudden changes.

Consider the circuit of Fig. 8-7, in which the closing of switch S results

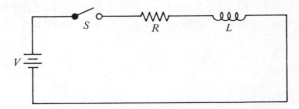

FIG. 8-7. A simple RL circuit which is energized from a dc-source by closing the switch S.

in a dc source voltage applied across a series resistance–inductance combination. As is well known, the current behavior following closing of the switch is as shown in Fig. 8-8a. Note that initially the current is zero and that finally it has the value V/R. This behavior is dictated by the fact that the current through an inductor cannot change by a finite value in zero time. Since the current is zero before the switch is closed, it must rise from zero by a smooth curve. Then, after all transients have disappeared, the final value of the current is limited only by the circuit resistance R.

The current through the inductor–resistor circuit may be considered to be the sum of two currents, as shown in Fig. 8-8b. The steady-state current,

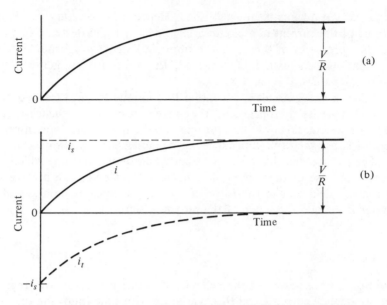

FIG. 8-8. The build-up of current in the circuit of Fig. 8-7. (a) The actual current. (b) The current decomposed into a steady state current i_s and a transient current i_t which, if added, give the total current i.

i_s, is assumed to be present from time $t = 0$, the start of the transient. A transient current, i_t, begins with a value $-i_s$ and decreases in value with time. Then

$$i = i_s + i_t$$

At $t = 0$,

$$i = 0$$

At $t = \infty$,

$$i = i_s = \frac{V}{R}$$

The same type of behavior applies in the presence of an ac source (Fig. 8-9). Now the steady-state current i_s is an alternating current (Fig. 8-10). The transient current i_t is of such magnitude that at $t = 0$, $i = i_s + i_t = 0$.

As in the dc case, let it be assumed that the steady-state current exists at full value from the instant the switch is closed (Fig. 8-10). Its value is determined by the source voltage and the impedance Z of the circuit. Note

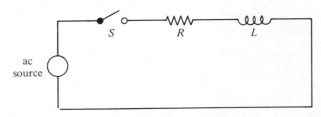

FIG. 8-9. A simple *RL* circuit which is energized from an ac-source

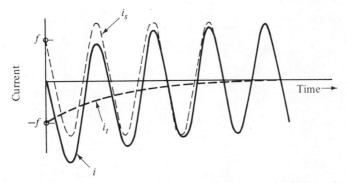

FIG. 8-10. The current in the circuit of Fig. 8-9 following the closing of switch *S*.

i_s-the steady state current.

i_t-the transient current.

i-the total current.

that in the example shown, the steady-state current i_s has a value of f at $t = 0$, the instant the switch is closed. Hence a transient current i_t of value $-f$ must be present. The transient current and the steady-state current are added together to give the actual current i, as shown.

When an ac generator is short-circuited on the occurrence of a fault, the resulting current contains both the steady-state ac component and the transient component. The magnitude of the transient component is dependent, as was shown in Fig. 8-10, on the point on the steady-state current wave where the fault happens to strike. The magnitude is greatest when the fault occurs at a point corresponding to the crest of the steady-state current; it is zero when the fault occurs at a point corresponding to the zero of the steady-state current. As can be seen from Fig. 8-10, the transient behavior of the circuit may cause the actual current to rise to a value substantially greater than that predicted by generator subtransient reactance considerations alone. To consider this effect, known as *dc offset*, some designers cal-

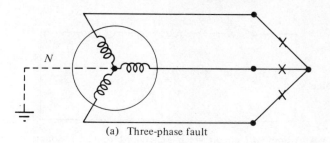

(a)　Three-phase fault

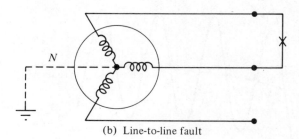

(b)　Line-to-line fault

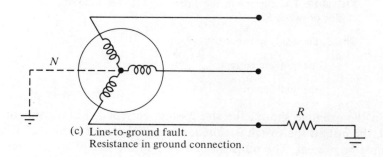

(c)　Line-to-ground fault.
　　 Resistance in ground connection.

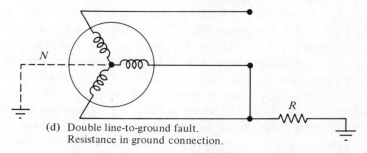

(d)　Double line-to-ground fault.
　　 Resistance in ground connection.

FIG. 8-11. Some typical types of faults on a 3-phase system.

culate the initial current of a short-circuited ac generator to be that current predicated by generator subtransient reactance X_d'', which is then increased by a factor of 1.2 to 1.7, depending on the ratio of the reactance to the resistance of the circuit.

8-6. UNSYMMETRICAL FAULTS ON THREE-PHASE SYSTEMS

As stated in Section 8-1, this discussion relates primarily to balanced three-phase faults. For these, all three conductors are assumed to be brought together into a short circuit simultaneously. As was described earlier, calculations for such faults may be made on a line-to-neutral basis or a single-phase equivalent basis.

Actually, the faults that occur on generators, transformers, and transmission and distribution lines are seldom of this type. Figure 8-11 shows a generator (neutral grounded) with faults of four different types, all of which must be considered in power-system design. A calculation procedure known as the *method of symmetrical components* provides a means for determining the currents and voltages to be expected at any point on a system on the occurrence of faults of the types illustrated in Fig. 8-11. The study of this subject is not discussed in detail here.

PROBLEMS

8-1. Determine typical values of X_d, X_d', and X_d'' (expressed in per unit and in ohms) applying to a 100,000-kVA 20-kV turbine-driven generator.

8-2. Refer to Problem 8-1. What will be the initial rms value of the 60-Hz current flowing with a three-phase fault at the terminals of the generator? Express in per unit and in amperes.

8-3. A power system consists of a 12-kV generator, a 138/12-kV transformer, and a 138-kV line. On a 100,000-kVA base these components have the following reactances:

generator $X_d = 2.40$ pu
$$X_d' = 0.30 \text{ pu}$$
$$X_d'' = 0.18 \text{ pu}$$

transformer
$$X = 0.26 \text{ pu}$$

line
$$X = 0.90 \text{ pu}$$

Consider a three-phase short circuit at the load end of the line. Calculate the current (in amperes) in the generator and the current in the line
a. immediately after the start of the fault.

    b.   10 cycles after the start of the fault.

    c.   2 sec after the start of the fault if breakers fail to clear.

Assume that the internal voltage of the generator is 1.1 pu.

8-4.   A 5-kV three-phase distribution circuit is supplied by a 200-kVA 34/5-kV transformer. The line is 4/0 ACSR spaced 7 ft equilaterally. Assume that the voltage on the primary side of the transformer is held constant at 34 kV. Solve for the current (in amperes) on the high-voltage side and on the low-voltage side of the transformer for a fault on the 5-kV line

    a.   at the station end.

    b.   4 miles distant from the station.

Neglect circuit resistance.

8-5.   Refer to Fig. 8-3. What will be the fault current after the opening of a circuit breaker in line B at bus I? What will be the voltage at bus II?

8-6.   Refer to Fig. 8-3. What will be the fault current if the fault occurs when line A is out of service? What will be the voltage at bus II?

8-7.   Refer to Fig. 8-4. Assume that each generator has an impedance of 1.0 pu and each line has an impedance of 0.5 pu. What is the fault current and what is the current in each line? Assume that each generator has an internal voltage of 1.2 pu.

8-8.   Refer to Fig. 8-4. Assume that the fault point is moved to a location at the middle of the center line. Outline a method for calculating the fault current.

8-9.   Two generators are operating on the same bus at 1.1 pu voltage. Other conditions are as follows:

	Generator 1	Generator 2
Reactance, pu	2.5	1.5
Load kW, pu	0.3	0.4
Power factor	1.0	0.8 lag

Solve for the internal generated voltage of each machine.

8-10.   A 100,000-kVA 20-kV ac generator is short-circuited at its terminals. What is the maximum possible crest value of the current, dc offset considered? $X_d'' = 0.11$ pu.

chapter 9

Circuit-Interrupting Devices

9-1. TYPES AND RATINGS

Circuit-interrupting devices in power systems are of many forms, ranging from the small key switches (in floor lamps and extension cords) to the enormous circuit breakers that control the flow of power in high-voltage networks. Three basic types of *circuit-interrupting devices* are recognized.

A *switch* is a device for making, breaking, or changing the connections in an electric circuit. This definition implies that the switch is used in a system in its normal operating condition.

A *circuit breaker* is a device for interrupting a circuit between separable contacts under normal or abnormal conditions. The abnormal conditions during which a circuit breaker may be asked to operate are primarily those which exist during short circuits. The circuit breaker is designed to withstand these short-duration, high-magnitude currents and to interrupt them if necessary.

A *fuse* is an overcurrent protective device with a circuit-opening fusible member directly heated and destroyed by the passage of overcurrent through it. Probably the most common application is in the service entrance boxes of residences, where fuses may be used to protect the main lines and each of the several individual circuits into the house. Fuses are also used in many applications on distribution primary circuits and to a limited extent on bulk-power high-voltage equipment.

All circuit-interrupting devices have certain design *ratings* that must be correlated with the duty to which they will be subjected. These ratings are as follows:

1. The power-frequency *voltage* of the circuit on which the device may be installed.
2. The maximum *continuous current* the device may carry.
3. The maximum *short-circuit current* the device is capable of interrupting.

The short-circuit volt-ampere rating of the device is defined from the rated circuit voltage and the rated short-circuit current:

$$\text{VA} = \sqrt{3}\ V_{\text{rated}}I_{\text{rated}} \tag{9-1}$$

The short-circuit ability is usually expressed in terms of kilovolt-amperes or megavolt-amperes. It is apparent that the short-circuit volt-ampere rating of an interrupting device provides no new information beyond that indicated by ratings 1 and 3.

Switches, circuit breakers, and fuses are of great importance in power-system design and operation. Reliance is put upon them for the changes of circuit configuration necessary as generators, transformers, and lines are placed in service or removed from service to accommodate changes in load conditions or necessary maintenance operations. In the event of abnormal conditions such as overloads or short circuits, circuit-interrupting devices must operate as quickly as possible. The high-speed isolation of a fault is of importance to minimize (1) damage to equipment at the point of fault, (2) the disturbance to the remainder of the system which might produce instability and loss of synchronism between machines, and (3) the deterioration of the circuit-interrupting device itself. Circuit breakers and fuses are available which interrupt short-circuit currents in less than two cycles of the power frequency.

In residential 120-volt lighting circuits, a switch is located in only one of the two conductors to lamps and small appliances. In power circuits, a circuit-interrupting device is put in *each* power conductor. These are sometimes termed three-pole circuit breakers (or fuses). The discussion that follows refers, for the most part, to a *single-pole device*, it being understood that three such units will be provided on a three-phase installation.

9-2. THE CIRCUIT-INTERRUPTING PROCESS

The important steps in the circuit-interrupting process are described with the aid of Fig. 9-1.

1. The circuit is complete with current flowing across the interface of two metallic bodies known as contacts (Fig. 9-1a).

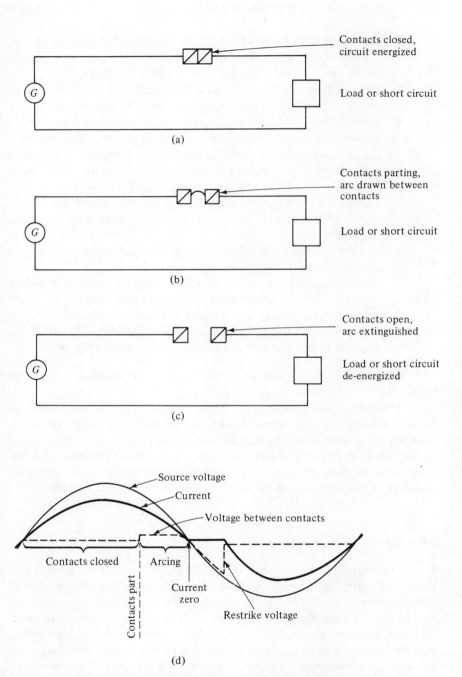

FIG. 9-1. Steps in the circuit opening process. (a) Contacts closed, circuit complete. (b) Contacts parted, circuit current flowing through an arc. (c) Contacts parted, arc extinguished, circuit open. (d) Current and voltage associated with the arc during the interruption process.

2. The contacts separate, and an arc is established between the parting contacts (Fig. 9-1b). The voltage across the arc is low, usually much less than the crest value of the circuit voltage.

3. In accordance with the normal cyclic changes in an ac circuit, the circuit current (and thus the arc current) will come to zero.

4. Immediately following the instant of current zero the voltage between the parting contacts will reverse in polarity. To reestablish the arc current in the opposite direction, a finite value of voltage is necessary. The restriking voltage will be greater than the voltage originally across the arc, as described in 2.

5. During the period of zero arc current (a finite time) the voltage between the arcing contacts rises at a rate depending on the nature of the circuit.

6. During this same period the voltage required to reignite the arc increases with time.

7. If the voltage of 5 is greater than the voltage described in 6 the arc restrikes, then continues until the next current zero to repeat the progress with greater gap spacing between the parting contacts.

8. If, on the other hand, the voltage of 6 is greater than that of 5 the arc fails to restrike and the circuit is interrupted (Fig. 9-1c).

Electrical conditions relating to arc behavior are diagrammed in Fig. 9-1d.

The design and operation of circuit breakers will be considered in Section 9-7 after the characteristics of arcs have been described. An understanding of the phenomena associated with arcs is helpful not only in regard to the operation of circuit-interrupting devices, but also in a study of lightning arresters, the sparkover of air across high-voltage insulators, and the puncture of solid insulation, as in generators, transformers, or cables.

9-3. CONDUCTION PROCESS

It is a well-accepted theory that all matter as we know it is composed of three basic particles: protons, neutrons, and electrons. The arrangement of these particles in nuclei, atoms, molecules, solids, liquids, and gases accounts for the physical and electrical properties of all materials as we observe them. A *proton* has mass and carries a unit *positive* electric charge. A *neutron* has a mass almost equal to that of a proton but carries no electric charge. An *electron* has a mass much smaller than that of a proton and carries a unit *negative* electric charge.

The *nucleus* of the hydrogen atom is made up of a single proton. The nuclei of atoms of other elements consist of a grouping of protons and neutrons. Each *atom* consists of a nucleus surrounded by one or more electrons.

Figure 9-2 shows models of atoms of three different kinds. In Fig. 9-2a an atom of hydrogen is shown with the nucleus made up of a single proton around which spins a single electron. In Fig. 9-2b the model of the

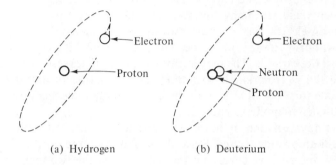

(a) Hydrogen   (b) Deuterium

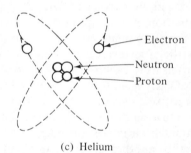

(c) Helium

FIG. 9-2. Models of simple atoms.

deuterium atom is shown with a nucleus consisting of one proton and one neutron. Again a single electron is shown in motion around the nucleus. Figure 9-2c shows the model of a helium atom. The nucleus consists of two protons and two neutrons, and around it move two electrons. Note that in each case in Fig. 9-2 the number of protons is equal to the number of electrons. As the chemical nature of the atom is determined by the number of protons, the deuterium atom and the hydrogen atom, each with one proton, have similar chemical properties but different mass. The helium atom, which has two protons, has chemical properties different from that of hydrogen. The atoms of other elements are similarly constructed, the number of protons in the nucleus being characteristic of the particular element.

All atoms in their normal state have protons and electrons of equal number. As each proton carries a unit positive charge and each electron carries a unit negative charge, it is obvious that each normal atom taken as a complete structure is *electrically neutral*. There are, however, many processes by which one or more electrons may be removed from a neutral atom. The electrons are then free negatively charged particles. The remainder of the atom, with more protons than electrons, has a net positive charge and is known as a *positive ion*. The removal of an electron from a neutral molecule is called *ionization* and the two particles produced are called an *ion pair*. A gas in which there are many free electrons and positive ions is said to be *ionized*. In an electric field between two oppositely charged plates, the free electrons will drift toward the anode. The positive ions, if free to move, will drift toward the cathode. The movement of these particles results in a transfer of positive charge from anode to cathode and a transfer of negative charge from cathode to anode. The sum of these two charge movements constitutes current flow from anode to cathode.

The removal of one or more electrons from a neutral atom requires energy from an outside source. Since the negative charge on the electron is attracted by the positive charge on the positive ion, the two tend to reunite in a process called *deionization*. Energy is given off when an electron unites with a positive ion. Ionizing processes and deionizing processes are of great importance in the formation of arcs and in their extinction.

Metals in solid form are made up of an assembly of a great number of atoms. A characteristic of metals is that as the individual atoms combine, an electron detaches itself from each of the many, many atoms forming the solid. The remainder of each atom forms a positive ion. These positive ions arrange themselves in a fixed, closely spaced, orderly array through which the free electrons may move as shown in Fig. 9-3a. The free electrons move randomly about within the structure traveling at high velocity from one boundary of the crystal to another. If an electric field is applied across the crystal, as shown in Fig. 9-3b, the electrons will drift from the negative terminal toward the positive terminal. This condition we call a current flow from the positive terminal to the negative terminal. The large number of electrons that drift freely in the presence of an electric field accounts for the high conductivity (low resistivity) of metals such as silver, copper, and aluminum.

Nonmetallic solids such as glass or porcelain consist of a somewhat ordered array of atoms of several different kinds. In materials that are regarded as insulators, there are very, very few electrons which are free to drift in the presence of an electric field. The structure of this matter is such that all electrons are kept in close association with their parent atoms.

Gases consist of individual atoms, or small ordered groups of atoms known as molecules, which are moving in random fashion at high velocity. They collide with each other and with the walls of the containing vessel. As the average separation between individual molecules is very great compared to their diameter, they may move relatively long distances between collisions. In the normal state each molecule is electrically neutral and so

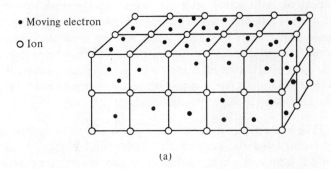

(a)

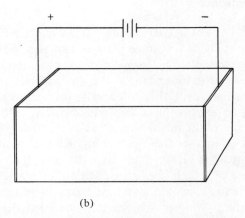

(b)

FIG. 9-3. Model of a metal. (a) The crystal structure showing fixed positive ions and moving electrons. (b) The metal has a voltage impressed across it.

has no tendency to drift in the presence of an electric field. In this situation a gas appears to be a perfect insulator. However, free electrons and positive ions introduced into the gas volume will render it conducting. The arcs used in arc welding and the arcs in circuit breakers during the circuit-interruption process are extreme examples of gases made conducting in this manner. The ionizing and deionizing processes that influence the conducting ability of a gas will now be discussed.

9-4. IONIZING PROCESSES

a. Radiation. All space on the earth is constantly bombarded and penetrated by very high energy submicroscopic particles. Some of these come from the sun and are known as *cosmic rays*. Others come from the disintegration of radioactive materials which are present in very minute quantities in the earth and even in all living organisms, including the human body. These high-energy particles, when passing near the electrons of a neutral molecule, may knock off one or more electrons from the molecule. With each such event, one or more free electrons are produced and one positive ion is generated. Because this process is going on continuously, it is found that all gases are slightly conductive.

b. **High Temperature.** As a gas is raised in temperature, the molecules constituting the gas move at higher and higher velocity. If the temperature is high enough, the collision of two gas particles may result in the detachment of one or more electrons and the generation of a positive ion. It is found that the gases constituting a high-temperature flame are highly conductive.

c. **Electron Collision.** If an electric field is set up in a gas volume by applying a voltage between two electrodes, those electrons and positive ions generated by cosmic rays and radiation particles begin drifting; the electrons move toward the anode, the positive ions toward the cathode. As these particles drift, they collide with neutral gas molecules. If the energy gained by an electron in a free path between collisions is sufficient, the collision may result in the ionization of another gas molecule, hence the generation of another free electron and another positive ion. Ionization by electron impact is of very great importance in the breakdown of gases. In the presence of a high-strength electric field a single electron moving from the cathode to the anode may have many ionizing collisions, each one of which produces another free electron which has the same ionizing ability. Such a multiplication of free electrons is known as an *electron avalanche*. Ionization by electron collision increases rapidly as the voltage gradient in the gas volume is increased.

d. **Emission From Metals.** Electrons may be released into a gas volume from a metal surface. Metals that have high boiling points, such as tungsten or carbon, when raised to a high temperature by any method, may "boil out" many electrons into the gas volume. This is known as *thermionic emission*. Another process, thought to be *high field emission*, may cause the discharge of electrons from a metal surface at a small area known as a *cathode spot* even at relatively low temperatures. Thermionic emission and high field emission are of great importance in all high-current arcs.

9-5. DEIONIZING PROCESSES

Deionizing processes are those means by which the number of free electrons and positive ions in a gas volume are reduced. In an arc, if the activity of the deionizing processes exceeds the activity of the ionizing processes, the number of free charged particles is reduced, the current in the arc diminishes, and arc extinction may result.

a. Electric Fields. The electric field between two plates of opposite potential continually removes charged particles from a gas volume, the free electrons moving toward the anode and the positive ions toward the cathode. At the anode the electrons enter into the metal, while at the cathode the positive ions each gain an electron from the metal surface to become a neutral gas molecule. If all ionizing processes were to cease, electric fields would quickly sweep all charged particles out of the gas volume.

b. Recombination. Recombination is the elimination of a free electron and a positive ion by their union to form a neutral gas molecule. Although in a highly ionized gas there may be many electrons and positive ions, recombination seldom occurs in the gas volume. It occurs readily at the surfaces of solids and liquids which are in intimate contact with the ionized gas.

c. Diffusion and Convection. Electrons and positive ions, in many respects, behave like ordinary gas molecules. They move from a region of high density by diffusion; they may be moved away from the region of ion-pair production by air currents.

d. Cooling. Cold air or other gases introduced into the highly ionized region of an arc does not eliminate charged particles directly but does reduce their number as it inhibits the formation of ion pairs due to high temperature collisions of gas molecules. Cooling is thought to be of great importance in the proper functioning of certain types of circuit breakers.

e. Electron Attachment. The molecules of electronegative gases tend to capture and hold free electrons which come close to them. The resulting particles, known as negative ions, in an electric field move toward the anode. However, because of their high mass, they do not gain sufficient energy between collisions with neutral gas molecules to produce ionization by collision. Hence the electrons trapped by this process are eliminated from producing new ion pairs.

The designers of electrical equipment use to their advantage the several processes of ionization and deionization. If a stable arc is required, as in an arc welder, conditions will be controlled to favor ionization and limit deionization. In a circuit breaker or in a lightning arrester, the problem is to eliminate the arc. Hence a design is chosen which favors deionization and discourages ionization.

9-6. ARCS AND ARC INTERRUPTION

Arcs may be started by several different methods. In the lightning flashover of transmission-line insulators, the extremely high voltage causes electric potential gradients in the air of such value that ionization by collision produces a condition known as sparkover. The spark quickly changes to the condition known as an arc as power-frequency current follows the breakdown path. In an electric welder system, the arc is started by separating two electrodes which are momentarily touched together. In a circuit breaker or fuse, the separation of the contacts results in an arc drawn between them. In relation to circuit-interrupting devices, we are not particularly interested in the mechanisms that start the arc. The characteristics of interest are those which may be used to bring about arc extinction.

A stable arc in a dc circuit is shown in Fig. 9-4. The current in the arc

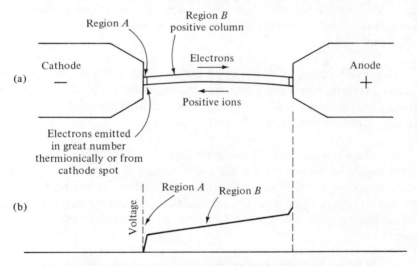

FIG. 9-4. Model of a dc arc. (a) Pictorial. (b) Voltage distribution.

may be a fraction of an ampere or many thousands of amperes. The cathode is of first importance in the arc because it emits electrons in great numbers. The movement of these electrons to the anode accounts for perhaps 90 percent of the current. These electrons may be emitted from the cathode due to thermionic emission if the cathode is of high-melting-point material such as carbon or tungsten. They may be emitted from a *cathode spot* by field emission if the cathode is of copper or other material of relatively low melting temperature.

The electrons move out from the cathode and in collision with neutral gas molecules in region *A*, (where the electric gradient is high) bring about *ionization by collision.* Now there are more electrons to move to the anode and some positive ions which move to the cathode. Region *B*, known as the *positive column*, is one of relatively uniform characteristics. The electric gradient in this region is quite low, but high enough to provide some ionization by electron collision. The new free electrons and positive ions are just enough to replace those lost laterally by diffusion out of the arc column and by recombination within it. At the anode, all current is due to the movement of electrons to it. These moving electrons cause some heating of the anode. The heavy positive ions arriving at the cathode at high velocities heat the cathode (if carbon) sufficiently to produce the high temperature necessary for thermal emission. On a copper cathode, their effect is to produce a cathode spot from which electrons are emitted.

The voltage across region *A* is 10 to 20 volts. The voltage across region *B*, the positive column, is proportional to its length, a few volts per centimeter.

If circuit conditions maintain the current constant while the spacing between anode and cathode is increased by the separation of the contacts, the voltage across the region *A* will remain substantially constant while the voltage across region *B*, the positive column, increases in proportion to its length. If the current is increased while spacing is held constant, slight changes of voltage across the arc will result in more copious emission from the cathode and more active ionization by collision in the positive column.

While the arc is maintained, deionizing agencies may be brought into play. Air may be blown across the positive column, carrying some of the ionized particles away while cooling the column. Oil may be sprayed into the arc or an insulating surface may be brought in contact with it, both providing increased recombination in the positive column. Electronegative gases may be introduced to capture some of the free electrons. All these processes require an increase in the gradient in the positive column if the arc is to continue to exist. In all circuits the gradient in the positive column is limited by the voltage which the circuit may provide between the arc terminals. Hence a point can be reached at which deionizing effects remove more charged particles than can be replaced by ionizing processes. In this case the arc will cease.

Theory indicates that an arc in a dc circuit of any voltage could be interrupted by a device that would introduce ample deionizing effects. However, practical limitations make it very difficult to design circuit breakers for use in dc circuits above a few thousand volts.

Arc interruption in an ac circuit is much more readily accomplished than in a dc circuit. In ac circuits, all currents and voltages go through cyclic changes. Hence ion-producing effects in the arc are variable, falling as current diminishes and ceasing at the time of current zero. Meanwhile, deionizing effects continue in full force. When the arc current comes to zero and the voltage across the contacts reverses in polarity, current in the new direction is not immediately established. For a short period of time the current remains zero. The length of the period of current zero is dependent on (1) the separation of the contacts, (2) the activity of the deionizing effects, and (3) the rate at which the voltage rises.

It may be noted that items 1 and 2 contribute to the suppression of the arc. This effect is sometimes described by saying that the arc path "gains dielectric strength."

If the arc path gains dielectric strength faster than the voltage across the arc terminals rises in value, the arc is not reestablished but ceases. If the voltage rises faster than the dielectric strength, arc current will flow in the new direction, at least until the next current zero. If the arc contacts are moving apart, hopefully a spacing is reached at which the arc is extinguished. It may be concluded that the extinction of an arc is dependent not only on the arc-interrupting device, but also on the characteristics of the entire circuit which controls the rate of rise of *recovery voltage* following arc current zero. Recovery voltage is discussed in Section 9-9.

The behavior of an arc regarding its extinction (as just described) applies to circuit-interrupting devices of *all* voltage classes. Another effect, of importance in arc extinction in circuits of relatively low voltage, is now described.

When a breaker is closed, there is high mechanical pressure between the breaker contacts, with the result that these pieces of metal touch each other

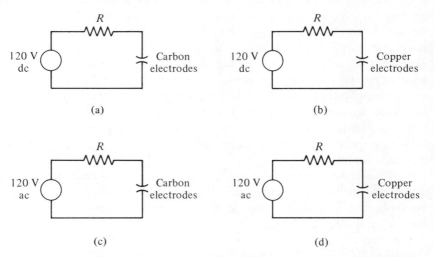

FIG. 9-5. An arc in a 120-volt circuit. (a) Dc-source, carbon electrodes. (b) Dc-source, copper electrodes. (c) Ac-source, carbon electrodes. (d) Ac-source, copper electrodes.

at many different points. As the pressure is removed, the number of points making contact reduces until eventually there is only one point of contact. Here the current density is very high and heating in a local area is very intense. On the final parting, a point on each contact is at very high temperature, a condition favorable to the emission of electrons from the electrode surface. If the contacts are made of high-melting-point material, electrons will be released by thermal emission and the temperature of the electrode maintained as positive ions move into it. When the potential between the electrodes reverses following current zero, thermal emission is picked up from the other electrode and the arc continues. However, the situation is quite different if the contacts are of low melting point, requiring that the electrons come from the metal by high field emission. The cathode spot does *not* hold over from one half-cycle to the next but must be reestablished following each current zero by phenomena within the arc itself. To do so requires a voltage across the arc, at least momentarily, of 300 to 400 volts. If the circuit is of such nature that it cannot provide this voltage, the arc ceases at the first current zero.

The behavior just described is easily illustrated by a set of experiments described in Fig. 9-5. Figure 9-5a shows a 120-volt dc circuit, a resistance R, and a pair of carbon electrodes. The resistance R may be of practically any value, perhaps 20 ohms. When the carbon contacts are parted, an arc is drawn between them, which continues until the electrodes are spaced a fair distance apart. Figure 9-5b shows a similar circuit but here the electrodes are copper. The behavior of this circuit is substantially the same as the one shown in Fig. 9-5a. Figure 9-5c is identical to the circuit of Fig. 9-5a except for the source, which is now an ac supply. When the electrodes are parted, again an arc is drawn, and its behavior is practically the same as that observed before. Figure 9-5d shows the same circuit except now the parting contacts are copper. As these contacts are pulled apart, there is a momentary flash but no continuing arc is established. Oscillographic records will show that the arc ceases with the first current zero. This behavior is explained by the fact that the 120-volt circuit has a crest value of 170 volts, substantially less than the 300 or 400 volts necessary to reestablish an electron supply by a cathode spot.

An oscillographic record of current and voltage in a circuit of Fig. 9-5c is shown as Fig. 9-6a. Note that only a small voltage, about 20 volts, is required to bring the thermionic cathode into full operation.

If the ac source voltage of Fig. 9-5d is increased, eventually a voltage is reached at which an arc can be reestablished between the copper electrodes following current zero. When this is done, an oscillographic record of current and voltage appears as shown in Fig. 9-6b. Notice that after current zero and voltage reversal, almost no current is observable until the voltage be-

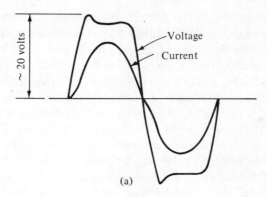

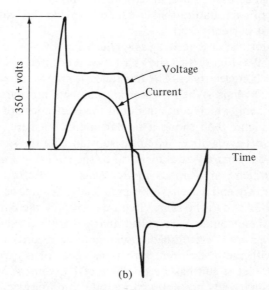

FIG. 9-6. The waveform of voltage and current in a short arc. (a) Between carbon electrodes. (b) Between electrodes of low melting point.

tween contacts has reached 350 volts. Then, with the reestablishment of a cathode spot, current increases and arc voltage decreases.

9-7. CIRCUIT-BREAKER DESIGNS

Designers of circuit breakers have resorted to many ingenious schemes to bring about arc interruption. Breakers for dc circuits rely principally on lengthening of the arc and subjecting it to deionizing effects. In breakers for operation in low-voltage ac circuits, advantage is taken of the voltage required to establish a cathode spot. In breakers for medium-voltage ac circuits, the cathode-spot phenomena may be combined with methods for bringing about the lengthening and deionization of the arc path. In breakers for high-voltage circuits, arcs are lengthened, forced against oil-laden surfaces, and cooled by oil sprays or by air jets. Examples of several types of breakers follow.

a. Air Circuit Breakers. The simplest method of extinguishing an arc is by increasing its length. Eventually a length is reached at which the voltage required across the arc is equal to the full circuit voltage. With further increase in length, the deionizing effects exceed the ionizing effects and the arc ceases. Figure 9-7 shows a simple arc-interrupting device with

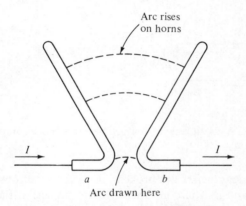

FIG. 9-7. A horn-gap arc-interrupting device.

which several arc-suppressing mechanisms may be illustrated. As the contacts *a* and *b* part, the arc is established at the base. Because of the heat generated, air currents cause the arc to rise. Furthermore, since electromagnetic forces in a loop tend to increase the area of the loop, the arc is further encouraged in its upward movement. The longer arc is readily cooled by convection air currents and by radiation, and the longer positive column requires a higher voltage. Each increase in length results in a higher voltage required to support the arc. Hopefully, in a circuit interrupter of this kind, the arc extinguishes before it reaches the top position on the contact horns.

Circuit breakers using the principles illustrated are frequently used in dc circuits and in low-voltage ac circuits. Coils carrying the *power current* may provide a magnetic field which aid in moving the arc toward the horn tips. Breakers of this type may be used to interrupt 100 amperes or more at several hundred volts.

The construction of a circuit breaker for use on a 120-volts ac circuit may be of very simple design. If the parting contacts are made of low-melting-point material such as brass or copper, the arc will go out at the first current zero as the voltage is insufficient to reestablish a cathode spot after current reversal. The necessary insulation must be so positioned as to avoid charring from the heat of the arc. If charring occurs, the carbon produced may serve as an electrode from which electrons may be released thermionically and so reignite the arc following current zero. Several hundred amperes may be interrupted by a simple breaker.

An arc interrupter for a circuit of higher voltage may be constructed as shown in Fig. 9-8. This is a combination of magnetic blowout to lengthen

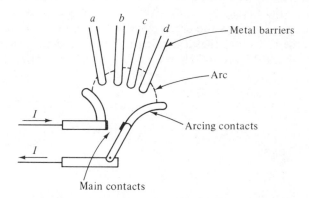

FIG. 9-8. A circuit breaker in which the arc is broken into several sections in series.

the arc, with the lengthened arc broken into several sections by the metallic barriers *a*, *b*, *c*, and *d*. There are then five arcs in series. Each one of these arcs experiences the effect of lengthening, cooling, and, if low-melting-point materials are used, the necessity of reestablishing a cathode spot following current reversal. Breakers of such design interrupt currents of several thousand amperes in ac circuits operating at several hundred volts.

Another circuit-interrupter design is illustrated in Fig. 9-9. Again the arc is lengthened by magnetic blowout. The barriers are insulating fins between which the arc is forced. Another factor, recombination and cooling at insulator surfaces, tends to reduce the ionization within the arc path and so increase the voltage necessary for restriking following current zero. This design may be used to interrupt currents of 50,000 amperes in ac circuits operating at 10,000 volts.

b. Oil Circuit Breakers. In many circuit breakers, arcs are drawn under oil (Fig. 9-10). As the arc plays between the parting contact, oil is vaporized to produce a large bubble, which surrounds the arc. Decomposition of the oil may result in this gas containing a considerable amount of hydrogen, a gas unfavorable to ion-pair production. This form of arc

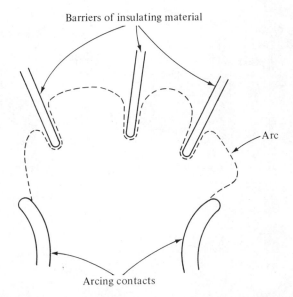

FIG. 9-9. A circuit breaker in which the arc is forced against transverse insulating barriers.

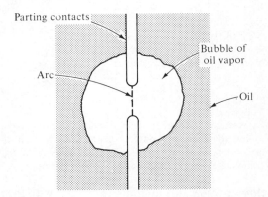

FIG. 9-10. An arc drawn under oil. A bubble of hydrogen surrounds the arc.

interruption, therefore, depends on the lengthening and cooling of the arc and the introduction of a gas unfavorable to arcing.

A more advanced form of breaker in which the arc is drawn under oil is shown in Fig. 9-11. Here an insulating chamber filled with oil surrounds the closed contacts (Fig. 9-11a). When the moving contact is lowered, the arc is drawn out through the throat of the insulating chamber (Fig. 9-11b). The portion of the arc within the chamber generates gas which forces oil

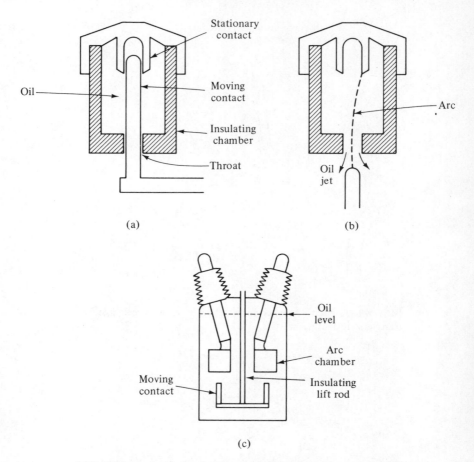

FIG. 9-11. A circuit breaker in which the arc is drawn under oil in a confining chamber. (a) Contacts closed, chamber full of oil. (b) Contacts open, arc drawn, oil being forced through arc region. (c) The complete breaker has two arc systems in series.

out through the throat, where it is in intimate contact with the arc to cool it, encourage recombination, and physically carry away the free electrons and positive ions. The arrangement of two sets of these parting contacts within a circuit breaker is shown in Fig. 9-11c.

Another version of the arc chamber in this type of breaker is shown in Fig. 9-12. Here the throat of the insulating chamber is made up of laminations between which oil may move radially into the arc path. In some designs, after the arc is drawn through the throat it is forced to move horizontally against barriers that are saturated with oil. Circuit breakers of this type have been built to interrupt 10,000-amperes short-circuit current in circuits operating at 500 kilovolts.

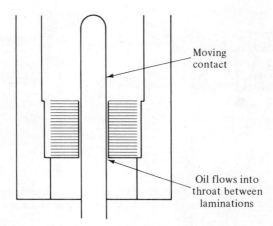

Moving
contact

Oil flows into
throat between
laminations

FIG. 9-12. An arc chamber in which oil is forced radially into the arc path.

Circuit breakers in which the contacts part under oil are very effective in performing their function as circuit interrupters. They have the disadvantage that the oil they contain presents a fire hazard in case the tank is ruptured due to unexpected internal pressure. Furthermore, the inertia of the heavy operating mechanisms limits the rate at which breaker contacts may part, with a consequent time delay in the opening of the arc. A 345-kilovolt oil-circuit breaker is shown in Fig. 1-16.

   c. Air-Blast Breakers. The air-blast breaker is a type in which the hazard of fire is reduced to a minimum and contact parting speed may be made very great. As the arc is drawn, a high-pressure air blast cools the arc and removes the ionized gas from between the parting electrodes. In Fig. 9-13a the air blast is parallel to the arc. In Fig. 9-13b the air blast is at right angles to the arc and so tends to extend it and force it against an insulating barrier, with the attendant recombination effect.

   A 138-kilovolts air-blast breaker is shown in Fig. 9-14a. In this breaker both the stationary and moving contacts are hollow tubes. As these contacts are parting, air under high pressure comes out of both and carries the arc products away from the contacts. On extra-high-voltage circuits, several air-blast units are connected in series to form a single circuit breaker. (See Fig. 9-14b.) Breakers capable of interrupting 40,000 amperes on 765-kilovolts ac circuits are in service.

   Vacuum switches (Fig. 9-15) provide another scheme for circuit interruption. In all our discussions so far we have considered the behavior of the gas between the parting contacts. As described earlier, this gas is ionized by several different processes and so provides free electrons, which move to

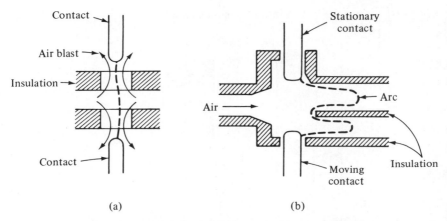

(a)   (b)

FIG. 9-13. An arc chamber in which an air blast is directed into the arc, (a) Parallel to the arc. (b) At right angles to the arc.

(a)

(b)

FIG. 9-14. An air-blast circuit breaker. (a) For 138-kV service. (b) For 345-kV operation, several units are connected in series.

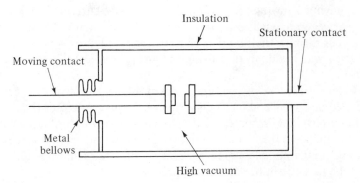

FIG. 9-15. A vacuum switch.

the anode, and positive ions, which move to the cathode. The positive ions coming to the cathode provide a mechanism by which electrons come out of the metal either thermionically or by high field emission. Practically all these phenomena disappear if the gas is removed from the space between the electrodes. The arc is interrupted at the first current zero and does not reignite. Vacuum breakers fail if leaks permit air to enter the interrupting chamber. The moving contacts are sealed by airtight metal bellows.

9-8. FUSES

A fuse, like a circuit breaker, is used to open a circuit in the event of an overload or short-circuit condition. Fuses are not ordinarily used to open and close circuits during normal operation. Unlike a circuit breaker, a fuse cannot be reclosed once it has opened. Its principle advantage is its low cost.

Fuses used on low-voltage circuits are of very simple construction (Fig. 9-16a). Two end terminals, 1 and 2, are joined together by a metal fuse

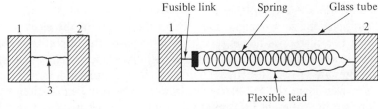

(a) Low-voltage fuse   (b) High-voltage, high-current fuse

FIG. 9-16. Fuses. (a) Cross-section of cartridge fuse. (b) High-voltage fuse.

wire, 3, having a low melting point. With excessive current through the fuse wire, heat generated causes it to melt and clear the circuit. At higher voltage and current ratings, the fuse wire may be surrounded by a powder (such as boric acid) which aids in deionizing the arc produced following melting of the fuse.

Figure 9-16b shows a design used on higher voltage. A short fusible link is held under tension by a coil spring. A flexible wire joins the link to the right-hand terminal. When excessive current causes the fusible link to melt, the spring pulls the open contacts apart, increasing the length of the arc. An insulating nonflammable liquid such as carbon tetrachloride may fill the entire tube to assist in the deionizing process.

9-9. RECOVERY VOLTAGE

As shown by previous discussion, interruption of high current arcs in circuit breakers of electric power systems is dependent on the fact that the power current passes through zero value and reverses direction twice during each cycle. As the instant of current zero is approached, those factors tending to maintain ionization within the arc are diminishing and momentarily are nonexistent. Meanwhile, those factors that tend to deionize the arc path are very active. Hence at the instant that the voltage across the arcing contacts reverses polarity, the stability of the arc is uncertain. Whether or not the arc is reestablished depends on the rate at which ionizing effects are brought back into play. Those phenomena which supply the free electrons and positive ions to the arc column are very voltage dependent. Hence the probability of arc reestablishment is increased by a high rate of rise of voltage across the arc terminal. One writer has described the problem of arc interruption following current zero as a race between two opposing sets of forces: the deionizing effects that rapidly increase the dielectric strength of the arc path, and the rapidly increasing voltage appearing between the parting contacts. Whether or not the arc restrikes following current zero depends on the outcome of this race.

The rate at which voltage rises across the contacts of a circuit breaker following current zero depends entirely on the nature of the electric circuit in which the current is flowing. A consideration of the circuit-interruption problem in some simple circuits illustrates the importance of *circuit configuration* on the rate of rise of recovery voltage following arc extinction at current zero. As seen from the following illustrations, the problem of circuit interruption is quite different in a resistive, an inductive, or a capacitive circuit.

a. Resistive Circuit. Refer to Fig. 9-17 which shows a circuit consisting of an ac power source whose terminals are 1-2, the contacts of a

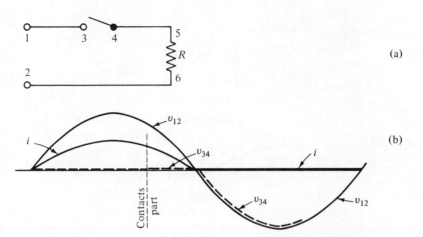

FIG. 9-17. Current interruption in a resistive ac circuit. (a) The circuit. (b) The current and voltage across the breaker contacts.

circuit breaker 3-4, and a resistor 5-6. From Kirchhoff's voltage law we may write

$$v_{12} = v_{34} + v_{56} \tag{9-2}$$

from which

$$v_{34} = v_{12} - v_{56} \tag{9-3}$$

When the contacts part, the voltage v_{34} is the voltage across the arc, perhaps less than 100 volts.

Assume that, following current zero, the circuit is interrupted; we may then write

$$i = 0 \qquad v_{56} = 0 \tag{9-4}$$

$$v_{34} = v_{12} \tag{9-5}$$

It is seen that in the resistive circuit, the voltage v_{34} available to cause restriking of the arc is exactly equal to the source voltage. Its rate of rise is determined from the characteristics of the source.

b. Capacitive Circuit. Figure 9-18(a) shows a capacitive circuit in which the parting contacts 3-4 are opening the circuit. Figure 9-18(b) shows the sequence of events following the parting of contacts and the interruption of current at the first current zero. From the relations

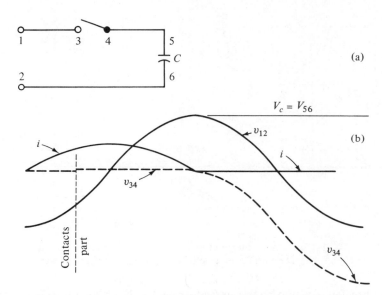

FIG. 9-18. Current interruption in a capacitive ac circuit. (a) The circuit. (b) The current and voltage across the breaker contacts.

$$v_{12} = v_{34} + v_{56}$$

$$v_{34} = v_{12} - v_{56}$$

before current zero and neglecting arc drop,

$$v_{34} = 0$$

$$v_{56} = v_{12}$$

Note that in the capacitive circuit, current zero occurs at the instant of the crest value of the voltage wave. This means, then, that if the current ceases at the instant of current zero, the capacitor is left charged with a voltage equal to the crest value of the ac supply, V_c. Then

$$v_{56} = V_c$$

$$v_{34} = v_{12} - V_c \qquad (9\text{-}6)$$

This equation shows that the voltage across the open contacts is the difference between the ac voltage and the dc voltage, as shown in Fig. 9-18b. Note that the initial rate of rise of voltage following current zero is relatively slow but that the voltage across the parting contacts at the end of one half-cycle reaches a value equal to twice the crest value of the supply wave. This high

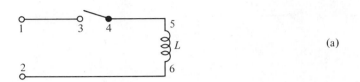

(a)

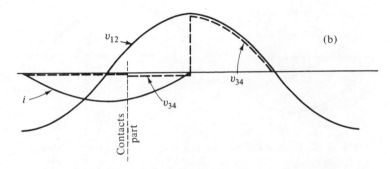

(b)

FIG. 9-19. Current interruption in an inductive ac circuit. (a) The circuit. (b) The current and voltage across the breaker contacts.

voltage across parting contacts may in an actual circuit breaker produce a delayed restrike of the arc. Such delayed restrike may give rise to serious transient conditions on the system.

    c. Inductive Circuit. A similar analysis on an inductive circuit is made with the aid of Fig. 9-19a and b:

$$v_{12} = v_{34} + v_{56}$$

Before contacts open,

$$v_{34} = 0$$

$$v_{56} = v_{12} = \frac{L\,di}{dt}$$

After contacts open,

$$i = 0 \tag{9-7}$$

$$\frac{L\,di}{dt} = 0 \tag{9-8}$$

$$v_{34} = v_{12} \tag{9-9}$$

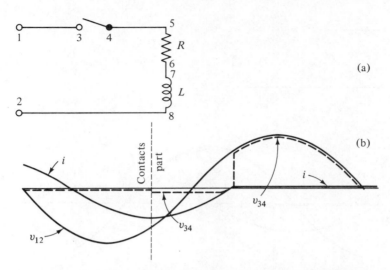

FIG. 9-20. Current interruption in a resistance–inductance circuit. (a) The circuit. (b) The current and voltage across the breaker contacts.

If the circuit current ceases at the instant of current zero, the rate of change of current through the inductor drops suddenly to zero and the voltage across the inductor becomes zero. Therefore, the voltage across the arc goes from zero value to the crest value of the supply voltage wave in zero time. This sudden rise of the recovery voltage tends to cause restrike of the arc.

d. Resistance–Inductance Circuit. A similar analysis of a series resistance–inductance circuit may be made with the aid of Fig. 9-20a and b. It is seen here that the presence of the resistance in the circuit shifts the point on the supply voltage wave at which current zero occurs and thereby reduces the recovery voltage appearing across the parting contact.

A similar analysis could be made on a resistance–capacitance circuit. Here again the recovery voltage would be found to be substantially reduced by the presence of the series resistors.

e. Resistance Switching. As shown in the discussion above, the presence of *series resistance* in a circuit reduces the rate of rise of recovery voltage across the parting contacts of the switch. In some applications a resistor is momentarily switched into the circuit to aid in circuit interruption, as shown in Fig. 9-21. With this arrangement the main contacts of the breaker 3-4 and the auxiliary contacts 3′-4′ are originally closed. When the circuit is to be interrupted, the main contacts 3-4 open first, leaving the resistor R' then in series with the remainder of the circuit. After a very short time

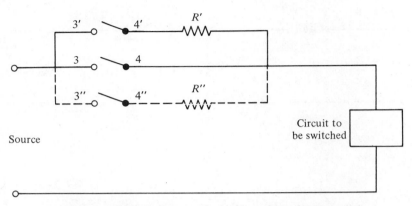

FIG. 9-21. Schematic diagram of a breaker with resistance switching.

      3-4   Main contacts.
      3'-4'   Auxiliary contacts for circuit opening.
      3"-4" Auxiliary contacts for circuit closing.

delay the auxiliary contacts 3'-4' open, completing the circuit interruption. This arrangement has a further advantage, in that it reduces transient over-voltages which may occur in the circuit to be interrupted. Transient over-voltages may also be generated when the circuit breaker is closed. These transients are reduced if the closure is made in two steps, the first bringing in the resistor as by the closing of contacts 3'-4' followed quickly by the closing of the main contacts, 3-4. In some switching arrangements it is advantageous to use a resistor on closing whose value is different from the resistor used in the opening operation. In that event another set of contacts 3"-4" and a resistor R'' (shown dashed in Fig. 9-21) are required. This circuit switching arrangement is particularly applicable to circuit breakers with which long transmission lines may be energized and deenergized.

     f. Inductance–Capacitance Switching. Figure 9-22a shows a circuit arrangement representing the inductance of a transformer through which is supplied a static capacitor, C. Suppose that a short circuit occurs on a connecting transmission line as shown and the circuit breaker 3-4 is required to remove this short circuit. It may be shown analytically (though it will not be done here) that following the interruption of the current at current zero, oscillatory voltage will appear across the contacts 3-4 (Fig. 9-22b). The (resonant) frequency of this oscillation is determined by

$$f = \frac{1}{2\pi\sqrt{LC}} \tag{9-10}$$

This type of transient results in a high rate of rise of recovery voltage and must be considered in circuit-breaker application.

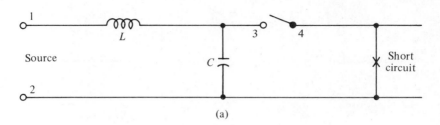

(a)

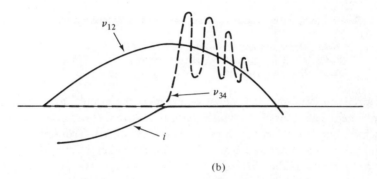

(b)

FIG. 9-22. Switching in the presence of series inductance and capacitance. (a) The circuit. (b) The current and voltage across the breaker contacts.

PROBLEMS

9-1.   A 100,000-kVA 18-kV three-phase ac generator is to be equipped with a breaker at its terminals. Specify the ratings of the breaker.

9-2.   Suppose the generator of Problem 9-1 is to be operated on the same bus with two other similar generators. How will its breaker requirement be affected?

9-3.   Describe the process of circuit interruption in a breaker on an ac system.

9-4.   What are the principal carriers of electric charge when current flows
a.   in a metal?
b.   in an arc?

9-5.   What is the basic difference between copper and porcelain which explains their vastly different conductivity (or resistivity)?

9-6.   List some of the things in nature that may produce free electrons.

9-7.   What is an ion pair?

9-8.   List and describe those processes in an arc which tend to maintain or increase the numbers of free electrons and positive ions present.

9-9.   List and describe those processes in an arc which tend to suppress the numbers of free electrons and positive ions present.

9-10.   List five mechanisms used by circuit-breaker designers in an effort to suppress arcs.

9-11.   Explain why it is much easier to interrupt an arc in an ac circuit than in a dc circuit.

9-12.   Compare the problem of interrupting 1000 A in a 120-V ac circuit with that of interrupting 100 A in a 1200-V ac circuit.

9-13.   What is meant by "recovery voltage"? What controls its rate of rise?

9-14.   Assume that a device were developed that would interrupt arc current at the instant of *crest* value. Make a sketch showing the rise of recovery voltage
        a. in a resistive circuit.
        b. in a capacitive circuit.
        c. in an inductive circuit.

9-15.   Refer to Fig. 9-22. Consider that L is the leakage inductance of a 25-kVA 22/5-kV transformer viewed on the 5-kV side. Let C be a 10-kVA capacitor bank for power-factor correction. What will be the frequency of oscillation when contacts 3-4 are opened?

chapter 10

System Instrumentation

10-1. SYSTEM MONITORING

A modern electric power system is an assembly of many components each of which influences the behavior of every other part. Proper functioning of the system as a whole makes it necessary to monitor conditions existing at many different points on the system in order to assure optimum operation.

The concern of the customers is primarily that the frequency and voltage of the supply are held within certain rather narrow limits. Since frequency of the system is the same everywhere, it may be monitored by a single frequency meter located at any convenient point. In contrast, the voltage of the system may be quite different at different points. Consequently, it is necessary to make continuous observation of the voltage at certain key points on the system in order to provide acceptable service.

Efficient operation of the system is obtained by assigning proper load schedules to each of the generators on the system. Newer plants, although individually more efficient, may be located at points on the system where their loading occasions large system losses. It is desirable to operate with a division of the load between generators such that the total cost of fuel consumed is minimized. To provide reliability of the power supply in the event of unexpected conditions, it is desirable to have the total kilowatt rating of all machines in operation somewhat greater than the total load plus losses.

This excess of generation, known as *spinning reserve*, is then available for picking up suddenly applied customer loads or to pick up the load dropped by a generator that must be removed from service for emergency maintenance.

Instrumentation is necessary to permit billing of customers for energy used. Many interconnections exist between different power systems. Instruments must be provided at interchange points to permit billing for energy transferred from one system to another. The continuous monitoring of energy transfer is necessary to assure that interchanged power is within the limits of contract agreements.

The continuous measurement of conditions on major pieces of equipment is necessary to avoid damage due to overload. As load increases from month to month, points at which additional capacity of equipment is required may be recognized and provision made for the installation of additional equipment. Thus instrumentation serves as a guide for future construction in a growing power system.

Occasionally, under emergency conditions, a system operator observes that his system load exceeds the ability of the available generating and transmission equipment. He is then faced with the problem of *load shedding* or, more properly, *load conservation*. It is then necessary to drop selected loads where service interruption is least objectionable. In such an event, he relies on the many instruments which provide information relative to system-operation conditions.

Instruments may sound alarms as advance warnings of conditions requiring action to avoid damage to equipment operating beyond its design limitation. In the event of extreme conditions such as power-system faults, defective equipment is switched out of service automatically. Instruments that continuously monitor current, voltage, and other quantities must be able to identify the faulted equipment and to bring about operation of the circuit breakers which remove it from service, while leaving in service all other equipment on the operating system.

The many different electrical devices on a power system and those owned by the customers are designed for operation within certain specified ranges. Operation outside these designed limits is undesirable, as it may result in inefficiency of operation, excessive deterioration, or (in extreme cases) the destruction of the device. Careful attention to the conditions under which equipment is operating may indicate corrective action that must be taken.

Overcurrent on all electrical devices is undesirable, as it produces excessive temperatures, inefficient operation, and reduced service life. Overcurrent in residential circuits may bring about disconnection of the circuit by fuse or breaker action. Overcurrent in motors may damage insulation, with possible early insulation failure.

Undervoltage considerably reduces the efficiency of incandescent lamps and may result in nonoperation of fluorescent lamps. Undervoltage of the power supply to motors may result in excessive currents in the motors, with possible damage to windings.

Overvoltage increases the light output of lamps but in many instances seriously shortens useful life. Overvoltage applied to motors and transformers may result in excessive losses within the iron, with possible damage to the iron or to the adjacent winding insulation.

Overspeed of rotating machines may result in structural damage to rotating parts. The overspeed of the customers' production equipment may result in an inferior quality of the product.

An out-of-step condition existing between two generators or between a generator and a synchronous motor results in an interruption of useful power transfer between the two machines. An out-of-step condition should be recognized promptly and the machines separated from each other. They may then be resynchronized and brought back into service.

Instruments of many different types must be installed at many locations on a power system and on the premises of the many customers. With such instruments, conditions existing on the system may be continuously monitored.

10-2. INSTRUMENTS FOR SYSTEM MONITORING

The instruments used for system monitoring include the following:

voltmeters	synchroscopes
ammeters	relays
wattmeters	varmeters
watt-hour meters	automatic oscillographs
frequency meters	

The meters may be *indicating* or *recording*. The instruments may in turn control automatic equipment for regulating voltage, frequency, power output, or power interchange on tie lines. In many modern systems, instrument readings are transmitted from many points on the operating system to control centers. At these control centers the reading of instruments from the remote points may be displayed as an aid to the system dispatcher. In some instances data from selected instruments are fed into a computer which determines plant loading for the most economic system operation. The computer then transmits signals to the various generators to load each machine to the predetermined value.

The transmission of instrument readings from point to point on a power system may be accomplished by several different methods. Leased or private telephone lines may provide circuits for information transmission.

Channels of communication may be provided by carrier current systems using the power conductors as the medium of transmission. In other instances, microwave circuits may be utilized. The system used for the transmission of instrument readings may also be used for voice transmission and for supervisory control.

10-3. INSTRUMENT CIRCUITS

Residental loads on electric power systems are ordinarily supplied by circuits that nominally operate at 120–240 volts. The currents that flow in these circuits may be as much as 100 or 150 amperes. The watt-hour meters which measure energy on these circuits have potential coils designed for operation at 120–240 volts and current coils which carry full-load current. In contrast, the watt-hour meters, wattmeters, voltmeters, ammeters, and other instruments used in power-system stations and substations and the instruments used on the systems of large industrial customers are, almost without exception, equipped with potential coils designed to operate at approximately 120 volts and with current coils designed to carry normal working currents of approximately 5 amperes. The 120-volt, 5-ampere line of instrumentation is accepted as standard. If voltages of power circuits, with values in some instances greater than 500 kilovolts, are to be measured by instruments with 120-volt coils, and if power-system currents measured in thousands of amperes are to be measured by instruments with 5-ampere current coils, it is necessary that some sort of voltage-sampling and current-sampling devices be interposed between the operating electric power system and the instruments that measure its performance. Potential transformers, capacitor potential devices, and bushing potential devices, connected to the high-voltage power system, supply to the instruments voltages that are a known fraction of the system voltage. Ideally the voltages supplied to the instruments are in phase with the system voltages which they represent. Similarly, current transformers and similar devices must supply instrument currents which are a known fraction of the power-system currents and which ideally are in phase with the power-system currents. Figure 10-1 shows a typical installation in which a single-phase load (or a line-to-neutral load) is supplied from an ac source. A potential transformer rated 72,000/120 volts provides a voltage supply for the potential coils of the voltmeter and the wattmeter. A current transformer rated 1000/5 amperes supplies current to the current coils of the instruments.

A consideration of Fig. 10-1 illustrates many advantages in the use of instrument transformers in power circuits. The power circuits themselves may be arranged in the switchyard almost independently of the location chosen for the instruments. Since the leads from the instrument transformers to the instruments themselves are subjected to voltages and currents of rela-

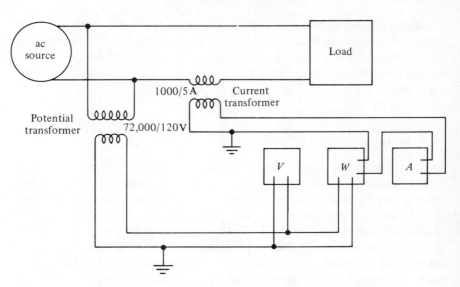

FIG. 10-1. A voltmeter, an ammeter, and a wattmeter connected to a power circuit through instrument transformers.

tively low magnitude, these connections may be constructed of wire of relatively small size and with minimum insulation. As a safety measure one side of each secondary circuit is grounded. The instruments may be located as desired in a control house, and the secondary leads run to them with relatively little hazard to the equipment or to the operating personnel.

10-4. VOLTAGE SUPPLY TO INSTRUMENTS

Several different types of devices may be connected to high-voltage circuits to supply the 120-volt circuits of the measuring instruments. All these systems have certain requirements in common. Ideally the voltage supplied to the instrument circuits should be a known fraction of the voltage on the power circuit and in phase with it. The device must insulate the power circuit from the instrument circuit, permitting one side of the instrument circuit to be grounded. If the voltage supplied by the instrument winding is exactly that fraction of the power-system voltage as indicated by the nameplate (such as 72,000 to 120) the device is said to have *zero ratio error*. If the output voltage is exactly in phase with the power-system voltage, the *phase-angle error* is said to be zero. Actually these ideal conditions are seldom attained in practice. The ratio and phase-angle errors may be of considerable concern, as they affect the accuracy with which the measuring instruments indicate the behavior of the power system itself. Ratio and phase-angle errors may change as the power-system voltage is changed. They are also influenced by the loading imposed on the potential transformer by the leads from it to the instruments and the instrument potential coils themselves. The load imposed on such a device is spoken of as the *burden*. The potential-reducing devices discussed here include the potential transformer, the capacitor potential device, and the bushing potential device.

The potential transformer (Fig. 1-24) provides the most accurate sampling of the power-system voltage of any of the available devices. It is also the most costly. The potential transformer is, in fact, merely a power transformer of full voltage rating designed for a very small secondary load, perhaps 200 volt-amperes. The equivalent circuit diagram of the transformer discussed in Section 2-8c may be used to analyze its performance. A transformer design that provides relatively low magnetizing current, low core-loss current, and low leakage reactance has extremely small ratio and phase-angle errors. Because of their high accuracy, voltage transformers are used almost exclusively to provide instrument potential in metering installations for billing to commercial customers or between interconnecting systems. Where installed for billing purposes, they are also used to supply the potential coils of indicating and recording instruments and relays.

Capacitor potential devices provide a less expensive method of obtaining a potential supply for instruments. However, greater phase-angle and ratio errors are inherent with this type of device. Their use, therefore, is limited to applications in which a high degree of accuracy is not required. Such devices are used to a considerable extent to supply potential to the voltmeters and wattmeters used to monitor electrical conditions within a system and to provide the voltage for the potential coils of protective relays. The capacitor potential device consists of a high-voltage capacitor shown in Fig. 10-2a contained in a porcelain weather shed associated with an impedance matching network (Fig. 10-2b). The impedance matching network as shown here consists of a transformer and adjustable resistors, inductors, and capacitors. Designs vary from one manufacturer to another but all provide a means for adjusting the output voltage to conform in ratio and phase angle to the high-side voltage. This adjustment may vary with the burden placed upon the device, and so adjustments may be required with the burden connected exactly as it will be operated.

Because the impedance-matching section is essentially a resonant circuit, ratio and phase-angle errors may change with applied voltage. Such changes may be particularly troublesome if there are nonlinear circuit elements associated with the secondary circuit, such as the iron-core potential circuits of relays. These ratio and phase-angle error changes may be of particular importance during fault conditions, when circuit voltage may drop to a small fraction of the normal value. It is, of course, during these

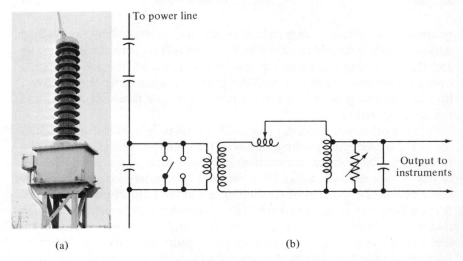

(a)   (b)

FIG. 10-2. A capacitor potential device. (a) The high voltage
capacitor. (b) The impedance matching network.

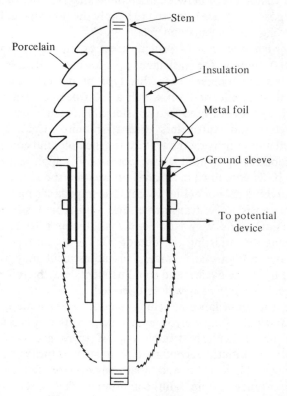

FIG. 10-3. The construction of a capacitor-type high-voltage
bushing.

times that relays must perform their function of identifying fault locations and removing the defective equipment.

A *bushing potential device* uses the capacitance of a high-voltage bushing in place of the high-voltage capacitor described in the previous paragraphs. As shown in Fig. 10-3, a high-voltage bushing is constructed of a central metal stem on which is wound a layer of insulating material. After building up this layer, perhaps a quarter of an inch, a layer of metal foil is put on. This process is repeated until the entire insulator system is assembled. In operation, a voltage appears on each of the metal layers, the lowest voltage appearing on that metal layer nearest the ground sleeve. The capacitance between this particular metal sleeve and the stem of the bushing serves as the capacitor which couples the instrument circuit to the high-voltage circuit. As the capacitance here is lower than that built into the high-voltage capacitors of the capacitor potential device, this device is less intimately connected to the power line. As a consequence, the ratio and phase-angle errors of bushing potential devices are considerably greater than those of capacitor potential devices. Hence they are used only in those locations where such error may be tolerated.

10-5. CURRENT SUPPLY TO INSTRUMENT

The current supply to switchboard instruments, relays, and similar equipment is most commonly provided by means of *current transformers.* Current transformers are iron-core devices with a primary winding consisting of a small number of turns of large current-carrying capacity. This winding is insulated from the core and the secondary for full power-circuit voltage. The secondary winding consists of a large number of turns of relatively small wire. The turns ratio is adjusted to supply 5 amperes to the instrument circuit when rated full-load current flows in the primary.

The current flowing in the primary of a current transformer is determined entirely by the power circuit in which the transformer is installed. This current is substantially independent of the characteristics of the transformer itself or of the secondary burden imposed upon it. However, for purposes of analysis it is convenient to treat the current transformer in the conventional fashion, beginning with the secondary current and adding to it the exciting current to determine the primary-current, turns ratio considered.

The equivalent circuit of the transformer (Fig. 10-4) serves to describe the performance of the current transformer. Ideally, the secondary of a current transformer should be short-circuited. Actually the impedance placed on it is very low, consisting of one or more instrument current coils and the connecting leads. Referring to the equivalent circuit, it is seen that if the

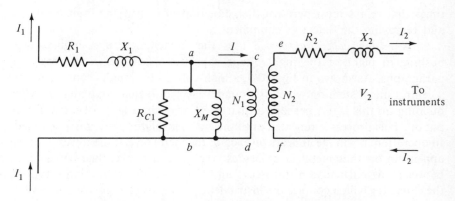

FIG. 10-4. The equivalent circuit of a current transformer.

transformer is short-circuited, the output voltage is zero and the voltage across the section *a-b* which accounts for the core loss and magnetizing currents is very low. If the transformer exciting current is so low that it may be neglected, it follows that the secondary current will be equal to the primary current multiplied by the turns ratio, N_1/N_2. Thus under the condition of zero burden on the transformer, the ratio and phase-angle errors will be very small.

If the connecting leads from the current transformer to the instruments are long and if several instrument current coils are placed in series, the transformer secondary load impedance is of importance and the burden on the transformer is said to be high. In order to circulate current in the secondary circuit to the instruments, a significant voltage must be present across the transformer secondary output terminals. To produce this voltage there must be a flux in the core, and to produce the flux there must be exciting current. As a consequence, the primary current, which is the sum of the load current and the exciting current, will be greater than the secondary-current, turns ratio considered. As a result a phase-angle and ratio error appears (Fig. 10-5).

The core and windings of a current transformer are relatively inexpensive. However, if the transformer is to be used for measuring the current on a high-voltage circuit, insulation and shielding for full line voltage must be provided between the primary and the core and between the primary and secondary windings. This results in a very substantial increase in the cost of a transformer. To minimize the insulation cost, bushing-type current transformers were developed. These transformers utilize the bushings which penetrate circuit breaker and power transformer cases as the insulation for the current transformers. A bushing-type current transformer is shown diagramatically in Fig. 10-6a. The power conductor whose current is to be measured penetrates once through a large-diameter ring-shaped laminated

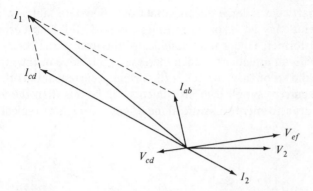

FIG. 10-5. A phasor diagram showing the voltages and currents associated with a current transformer.

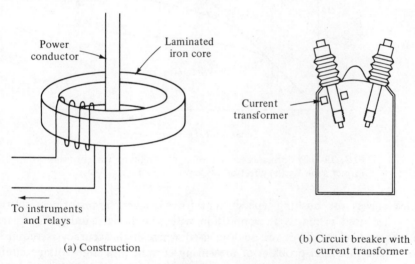

(a) Construction

(b) Circuit breaker with current transformer

FIG. 10-6. A bushing type current transformer. (a) Basic structure. (b) Installation in a circuit breaker.

iron core. The secondary winding on this core connects to the associated instruments and relays.

Figure 10-6b shows the cross section of an oil circuit breaker with a current transformer in place. It may be noted that the breaker bushing provides the insulation between the power conductor and the secondary winding.

As may be seen from Fig. 10-6a, the bushing-type current transformer consists of a single primary turn rather loosely linked with the secondary. As a result, the transformer leakage reactance is relatively high. This gives rise to increased ratio and phase-angle errors inherent in the bushing-type current transformer as compared with the ones of conventional design.

Errors are particularly large during conditions of system faults, when short-circuit currents may be 10 or more times the rated full-load current of the current transformer. Under fault conditions, the iron of the core may reach saturation with an attendant sudden increase in the value of the magnetizing current. As the secondary current is the primary current minus the exciting current, the current supplied to instruments will be less than the value indicated by the transformer turns ratio alone. Figure 10-7 shows typical perform-

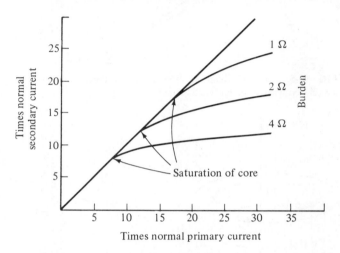

FIG. 10-7. Performance curves of a bushing-type current-transformer showing the effect of burden.

ance curves for bushing-type current transformers. Transformers of this type are used primarily in connection with switchboard indicating instruments and relays. They are seldom used where high accuracy is required.

The principal problem of measuring current in a high-voltage circuit lies not in the current measurement itself but in the transfer of the information to a circuit which may be conveniently connected to switchboard equipment. For example, a relatively inexpensive ammeter could be connected directly into a power conductor. Its reading might be observed by looking through a telescope. This arrangement would, of course, not provide circuitry for switchboard equipment. Several schemes have been developed for measuring the current with the measuring equipment itself tied directly to and operating at the same voltage as the power equipment. The current readings are then transmitted over a radio channel or over a light beam to a low-voltage circuit leading to switchboard equipment. These systems permit the simplest form of high-voltage insulation between the measuring system and the instrument circuits (perhaps a single column of porcelain insulators) but require the complexity of the information transmission circuitry. However, this type of current-measuring system is particularly attractive on extra-high-voltage circuits, where circuit-breaker design is of such form (see Section 9-7c) that no high-voltage entrance bushings are required.

PROBLEMS

10-1. A 72,000/120-volt potential transformer on a high-voltage line supplies a switchboard voltmeter. If the voltage on the power system is 68,000 volts, what is the reading of the voltmeter?

10-2. A 5000/120-volt potential transformer supplies a switchboard voltmeter. If the voltmeter reads 130 volts, what is the voltage on the power system?

10-3. A current transformer, rated 2000/5 amperes, is on a power conductor and supplies a switchboard ammeter. If the ammeter reads 2.1 amperes, what is the current in the power conductor?

10-4. The current in a power conductor is 650 amperes. A current transformer rated 400/5 amperes supplies a switchboard ammeter. What is the reading of the ammeter?

10-5. A single-phase power line carries a current of 350 amperes at 69,000 volts, 0.8 power factor lag. A switchboard wattmeter is supplied by a 72,000/120-volt potential transformer and 400/5 amperes current transformer. What is the reading of the wattmeter?

10-6. A wattmeter having a full scale reading of 500 watts connects to a power circuit through a 1000/5-amperes current transformer and a 138,000/120-volts potential transformer. The wattmeter is rescaled to indicate power-circuit quantities directly. What will then be the full scale reading of the wattmeter?

10-7. A bushing potential device having a nominal rating of 500,000/120 volts is improperly adjusted, with the result that the output voltage to the switchboard instrument is 8 percent low. With the switchboard instrument reading 104 volts, what is the voltage on the line?

10-8. A bushing potential device rated 72,000/120 volts is properly adjusted for ratio but has an improper phase-angle adjustment, which results in the secondary voltage being 10 degrees behind the primary voltage. This device is connected to a single-phase power line to provide potential for a switchboard wattmeter. Current to the wattmeter is supplied by a 200/5-amperes current transformer. On the line the voltage is 69,000 volts, the current is 180 amperes, and the power factor is 1.0. What is the actual power on the line, and what is the power as determined from the switchboard wattmeter?

10-9. Repeat Problem 10-8 but assume that the power factor is 0.5 lead.

10-10. A bushing-type current transformer is rated 400/5 amperes. During a fault, the short-circuit current is 8000 amperes. What current flows in the switchboard instruments if the impedance of the instruments and connecting leads is
a. 1 ohm?
b. 4 ohms?

chapter 11

Relays and
Relay Systems

11-1. GENERAL

A relay is a device that responds to variations in the conditions in one electric circuit to affect the operation of other devices in the same or in another electric circuit. The earliest forms of relays were electromechanical devices, a type that gained general acceptance because of its simplicity and rugged construction. Power-system relays must be highly reliable, for they may go for long periods without attention but must be ready to operate with speed and accuracy on demand. Most relays in service on electric power systems today are of this type.

Relays and relay systems using vacuum tubes were developed but were never generally accepted. The inherent uncertainties of vacuum tubes (which sometimes burned out or experienced a change in performance characteristics) made them somewhat unsuited for power-system protection.

Solid-state, or static, relays have, in the past few years, been developed to the point where they are highly acceptable for power-system protection. Circuit designs have been developed in which operating characteristics are almost entirely independent of those variations of component characteristics that result from temperature changes or age. The response time of solid-state relays may be made as short as desired; in fact, the basic components themselves have such fast response that it is frequently necessary to add delay circuits to avoid false operations due to momentary transients on the power

247

circuits. The various functions necessary in relay applications can each be accomplished by solid-state circuitry of different designs. It is, therefore, quite difficult to show solid-state circuitry basic to a particular relay function. However, certain examples are presented in Section 11-11.

The principles of protection of electric power circuits are quite independent of the relay designs which may be applied. For example, if the current to an electric circuit or a machine is greater than that which can be tolerated, it is necessary to take remedial action. The device for recognizing the condition and initiating corrective measures would be termed an over-current relay regardless of the mechanisms by which the function would be accomplished. Because the functions of electromechanical devices are easily described, their performance will serve as a basis for presenting a description of relays and relay systems in general.

11-2. BASIC RELAY TYPES

Most of the electromechanical relays in use may be classified into a few basic types. Each of these has moving contacts that may be made to open one or more circuits, to close one or more circuits, or to do both. Each of these devices is ordinarily in the normal state and moves into the operate state on the occurrence of certain electrical changes in the coils of the relays.

a. Moving-Armature Relay. The *moving-armature relay* is shown in Fig. 11-1a. A coil C is wound on a core of magnetic material M. A movable armature A carries with it a movable contact B. With no current in the coil, the contact B rests against the contacts 1-1, connecting them together. Gravity or a spring holds the armature in the position shown. If current through the coil is slowly increased, a point is reached at which the armature A suddenly rotates to the left, opening contacts 1-1 and closing contacts 2-2. The minimum current at which this motion may be initiated is termed the *pickup current*. If the current is now slowly reduced, a value will be found at which the armature will drop back to its original position. This is termed the *drop-out*, or *release*, *value*.

b. Solenoid. A *solenoid relay* is shown in Fig. 11-1b. A fixed coil surrounds a movable iron plunger. The plunger is held away from a symmetrical position in the coil by gravity or by a spring. When current of sufficient value flows in the coil, the iron core is drawn into the coil, thus closing the contacts as shown. Obviously such an arrangement could be made to open contacts by a simple position change. Again the pickup current is greater than the release current. The plunger in any selected position experi-

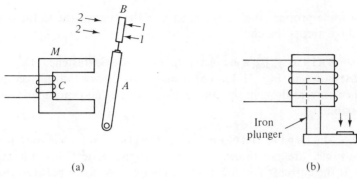

(a)

(b)

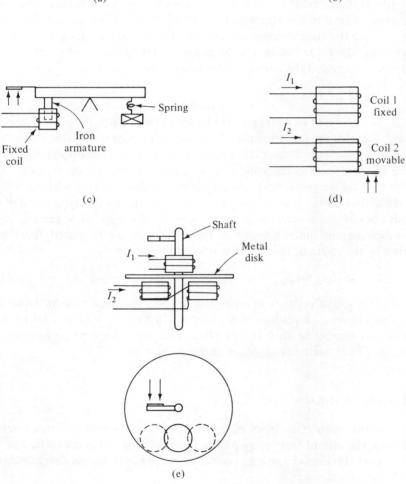

(c)

(d)

(e)

FIG. 11-1. Electro-mechanical relay types. (a) Moving armature.
(b) Solenoid. (c) Balance beam. (d) Mutually interacting coils.
(e) Induction disc, elevation and plan.

249

ences a force proportional to the square of the current (or to the square of the voltage across the coil).

c. Balance Beam. An iron armature is fastened to a *balance beam* as shown in Fig. 11-1c. The armature is drawn into a fixed-position coil when current flows in the coil. Various contact and spring arrangements are possible.

d. Mutually Interacting Coils. Two coils, one fixed and one movable, are positioned so that their magnetic fields interact (Fig. 11-1d). If the currents are of constant direction, the force between them will be proportional to the algebraic product of the two currents and may result in drawing the coils together or thrusting them apart. If ac currents flow in the coils, the force action will be proportional to the product of the two current values and the cosine of the angle Δ between them:

$$F = K_M I_1 I_2 \cos \Delta \qquad (11\text{-}1)$$

e. Induction Disc or Induction Cup. The *induction-disc relay* operates on the principle of the induction motor. As illustrated with Fig. 11-1e, currents displaced in time flowing through coils displaced in space give rise to a moving magnetic field which passes through the movable disc. This moving magnetic field sets up within the disc currents which result in force action on the disc, tending to make it rotate. In this type of relay the coils above the disc may carry one current and the coils below the disc carry another current independently. If these currents are displaced from each other by the angle Δ, the torque produced on the disc is

$$T = K_D I_1 I_2 \sin \Delta \qquad (11\text{-}2)$$

If the two sets of coils are in series and so carry the same current, no torque results. However, a resistor or a capacitor in parallel with one set of coils causes its current to shift out of phase with the other and so give rise to torque. Other methods of phase shifting are used.

11-3. relay timing

Any of the basic relay types may be modified to control the time interval between the instant from relay pickup to the closing of the contacts. For the most part these modifications consist of mechanical accessories associated with the relays.

a. Instantaneous. An *instantaneous relay* is one in which no intentional time delay has been provided. Figure 11-2a represents an instan-

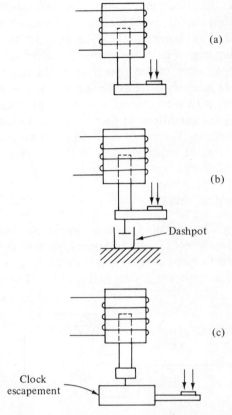

FIG. 11-2. Relay timing mechanisms. (a) Instantaneous. (b) Inverse time. (c) Timed.

taneous solenoid type of relay. Certainly there will be a short time interval between the instant of pickup and the opening or closing of the contact, but no intentional time delay has been added.

b. Inverse Time. An *inverse-time solenoid relay* is shown in Fig. 11-2b. A dashpot (either hydraulic or pneumatic) attached to the moving plunger slows its upward motion. At a current value just equal to pickup the plunger moves slowly and the time delay is at a maximum. At higher values of coil current, the pull on the plunger is greater and the time delay is shortened. Each increase in current shortens the time delay, hence the name, inverse time relay.

In the induction-disc relay, inverse time delay may be introduced by positioning a permanent magnet so that its field penetrates the disc. When the disc moves, currents set up in it produce a drag on the disc, which slows its motion.

   c. **T i m e d .** In some applications it is desirable to have a definite time interval between the instant of pickup and the closing of contacts. This may be accomplished as shown in Fig. 11-2c, in which the movement of the plunger of the relay trips a clock escapement mechanism. This mechanism, after a definite preset time, causes operation of the relay contacts.

   Various other schemes of relay timing are in use. In some instances relays are equipped with mechanisms to control the time interval between the instant of dropout conditions in the coil to the restoration of contacts in their normal position. The use of time delays in both trip and restore is illustrated in some of the examples later in the chapter.

11-4. RELAY-SYSTEM EXAMPLES

Several examples will be presented to show some of the principles of applications of relays to the control of circuits. Special types of relays to be discussed later will be applied to more complicated systems.

   A simple relay system is shown in Fig. 11-3. The output of a light-

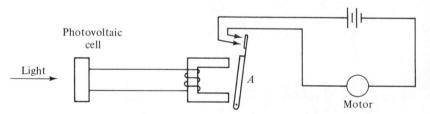

FIG. 11-3. A relay system by which a photo-cell controls the operation of a motor.

sensitive photovoltaic cell is connected to the coil of a moving-armature relay. When light strikes the cell the relay picks up, closing contacts and completing the circuits from the battery through the motor. If the light decreases to such a value that the current in the coil is below the release value, the armature returns to the normal position opening the contacts of the relay and deenergizing the motor. It is evident that by replacing the relay with one in which the contacts were closed in the normal position and open in the operate position, the sequence of events could be reversed. With light on the cell the motor is shut down and runs only when the cell is dark.

   Figure 11-4 is a modification of Fig. 11-3, in which a solenoid type of relay has been added in a connection by which it serves as a seal-in relay. Note that when relay *A* closes contacts, the current from the battery to the motor must flow through the operating coil of relay *B*. When its contacts close, it short-circuits the contacts of relay *A*, making further operation of

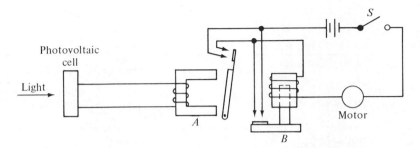

FIG. 11-4. A system similar to that of Fig. 11-3 but with the addition of a seal-in-relay.

that relay of no consequence. Relay B will remain in the contacts closed position until switch S is opened, interrupting the current to the motor and to the coil of relay B. Seal-in relays have many applications in the electric power field.

A third example of relay applications is shown in Fig. 11-5. Three independent circuits may be identified in this diagram. The first is a power circuit connecting the generator to the load. A circuit breaker provides a means for opening this circuit. A current transformer provides a small sample of the power current to the relay coil, I. A potential transformer provides a small sample of the potential of the power circuit to the coil of relay II. The movable contacts of the circuit breaker are carried on an insulating operating rod, which in turn is provided with other devices. A second circuit provides a connection between battery X and a solenoid-operated latch on the operating rod. In normal conditions with the circuit breaker closed, relays I and II are in the contacts-open position, while the auxiliary contacts on the operating rod are closed. In parallel with the trip coil of the latch is another solenoid relay, III, whose contacts are normally closed. A third circuit provides a connection through a manually operated closing switch T to the circuit-breaker closing solenoid.

If the breaker is open, it may be closed by momentarily closing switch T, thus energizing the closing solenoid and lifting the operating rod into the latch position, simultaneously closing the auxiliary contacts. The breaker may be opened manually by closing switch S, which connects battery X to the trip coil and triggers the latch, letting the breaker fall open.

If the breaker is closed, it may be opened automatically on the occurrence of abnormal power circuit conditions. If the current is too high, the over-current relay I will pick up, closing contacts and thus energizing the trip coil. Similarly, if the breaker is closed and the voltage of the circuit goes below a predetermined value, the undervoltage relay II will drop out, closing contacts and energizing the trip coil. In each case, when the trip coil is energized and the operating rod moves downward, the auxiliary contacts open,

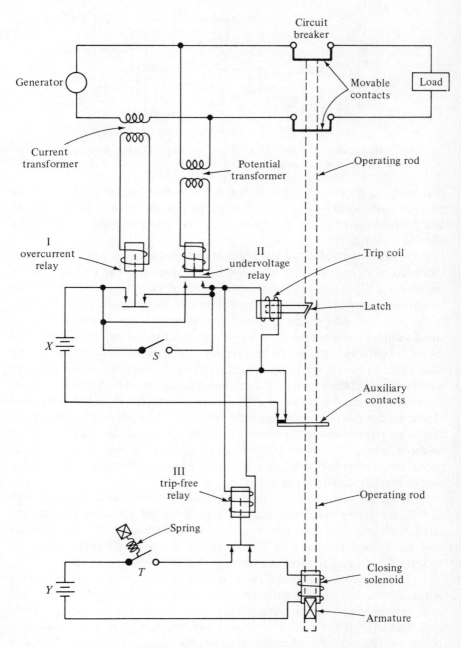

FIG. 11-5. A system for opening and closing a circuit breaker.

thereby opening the trip-coil circuit. In the event of a short circuit on the load side of the breaker, circuit conditions would cause the overcurrent relay I to pick up and the undervoltage relay II to drop out, either of which could complete the circuit to the trip coil, thereby opening the breaker.

Note that circuit-opening relay III has its coil in parallel with the trip coil. This is to provide trip-free operation of the breaker. Suppose the breaker has opened under short-circuit conditions and that this short circuit still exists. If the breaker is closed by hand by closing switch T, the presence of the fault will be detected by relays I and II, which will provide current to the trip coil and to the coil of the *trip-free relay*. The trip-free relay, III, on operating will open contacts and so cut off current to the closing solenoid. A time delay on the restore function of the trip-free relay would provide opportunity for the station operator to take his hand off switch T before relay III falls back into closed position.

In summary, the circuits of Fig. 11-5 provide control over the power circuit from the generator to the load. This circuit can be opened and closed manually. It will be opened automatically on the occurrence of overcurrent, undervoltage, or both. If the breaker is closed manually in the presence of a fault on the load side of the breaker, it will promptly reopen.

11-5. FUNCTIONAL RELAY TYPES

Basic relay types were presented in Section 11-2. It may be noted that in each of these relays a single force exists to cause operation of the relay. In more advanced types, several different forces may operate simultaneously on the moving element of the relay. These more complicated relays may serve special functions in the protection of electric power systems. Most of these special-function relays are produced by modifying either of two types of basic relays: the balance beam or the induction disc. The discussion presented here refers to the balance-beam type of relay.

Three types of coil arrangements may be used to influence the beam position. The first is a fixed coil wound with many turns of wire across which normal circuit voltage might be impressed, which will be called a voltage coil. This coil, when energized, produces a force on an armature attached to the beam. The second, a fixed coil with a small number of turns of relatively large wire designed to carry continuously a current of at least 5 amperes, will be designated as a current coil. The current coil will operate on an armature attached to the balance beam. The third type of arrangement, designated the watt element, consists of a fixed current coil positioned to interact with a movable potential coil attached to the beam. A voltage coil, a current coil, and a watt element may be used individually or in combination on a single balance beam. In some cases a spring will be attached to the balance beam to provide a restraining force.

a. Voltage Coil and Spring. A balance beam with a voltage coil and spring is shown in Fig. 11-6. The torque that is set up electrically, which tends to tip the beam, is

$$T_v = K_v V^2 \tag{11-3}$$

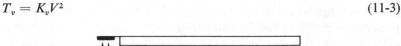
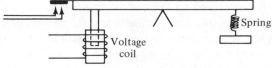

FIG. 11-6. A balance beam relay, voltage controlled.

where K_v depends upon the construction of the coil, the armature, and the length of the lever arm. When the beam is in the equilibrium condition, the torque produced by the spring T_s is exactly equal to the electric torque, or

$$K_v V^2 = T_s \tag{11-4}$$

As a starting point, assume that with normal voltage on the coil the forces on the beam are exactly balanced. Now if the spring tension is increased, the beam tips clockwise and remains in this position with normal voltage. However, if the voltage is increased, a point will be reached where the electric torque exceeds the spring torque and the beam tips counterclockwise, closing the relay contacts. This then provides the structure of a relay for detecting *overvoltage*.

Beginning again with normal voltage and with the beam in the balanced position, let the spring tension be decreased until the beam tips counterclockwise. It remains counterclockwise as long as the voltage is of normal value. However, if the voltage is reduced, a point is reached at which the beam rotates in a clockwise direction. This, then, is a structure by which *undervoltage* may be detected.

b. Current Coil and Spring. A balance beam with a current coil and spring is shown in Fig. 11-7. The torque produced by the current coil T_i is

$$T_i = K_i I^2 \tag{11-5}$$

The electric torque is equal to the spring torque T_s and the beam is balanced when

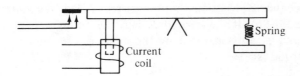

FIG. 11-7. A balance beam relay, current controlled.

$$K_i I^2 = T_s \tag{11-6}$$

By the same arguments as were presented in Section 11-5a, it may be shown that by proper adjustment of the spring, the relay of Fig. 11-7 may become either an overcurrent relay or an undercurrent relay.

c. Watt Element and Spring. A balance beam with two mutually interacting coils and a spring is shown in Fig. 11-8a. The electrically generated torque was given by Eq. (11-1) as

$$T = K_M I_1 I_2 \cos \Delta \tag{11-1}$$

In the application to be considered here, coil 1 may be made of many turns of fine wire and be connected across a voltage supply V through a series impedance Z_s. The current I_1 will then lag the voltage V by some angle, depending on the total impedance of the series circuit made up of the coil

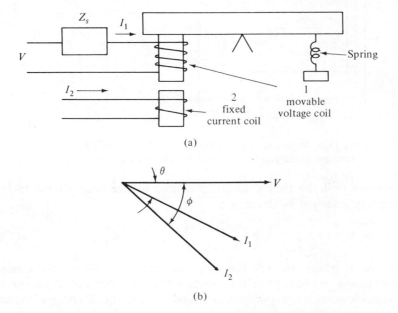

(a)

(b)

FIG. 11-8. A balance beam relay with mutually interacting coils.

and the impedance Z_s. Assume that the current I_1 lags the voltage V by the angle θ (Fig. 11-8b). If the current I_2 lags the voltage V by the angle ϕ, currents I_1 and I_2 are separated by the angle $\phi - \theta$, and the electrically generated torque is

$$T_w = K_w V I_2 \cos (\phi - \theta) \tag{11-7}$$

When the beam is balanced, the electric torque is equal to the spring torque

$$K_w V I_2 \cos (\phi - \theta) = T_s$$

Again by proper adjustment of the spring, the relay can be made to operate on either an excess value or a deficiency of the electrically generated torque. The voltage coil may be connected to any potential terminals and the current coil may be connected into any current-carrying conductor. The electric torque nevertheless will be as given by Eq. (11-7). Two mutually interacting coils with this arrangement are commonly termed a *watt element*.

   If the current and voltage coils of a watt element are connected to a power circuit as shown in Fig. 11-9, in which Z_s is chosen of such a value

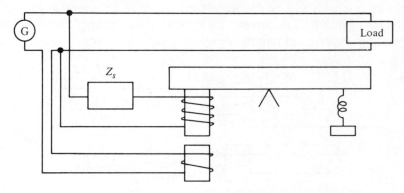

FIG. 11-9. A balance beam relay connected to respond to the interaction of the current and voltage in a power circuit.

as to make $\theta = 0$, the electric torque produced will be proportional to the power being delivered by the circuit:

$$T_w = K_w V I \cos \phi = K_w P \tag{11-8}$$

If the flow of power reverses, the electrically generated force on the relay will reverse and will be in the same direction as the spring force. If the spring force is made zero, the relay becomes an indicator of the direction of power flow.

d. Voltage Coil and Current Coil. Figure 11-10 shows a balance beam with a voltage coil and a current coil generating opposing torques. Any spring torque, if used at all, will be very small and may be neglected. The beam is balanced when the torque produced by the current coil is equal to the torque produced by the voltage coil, or

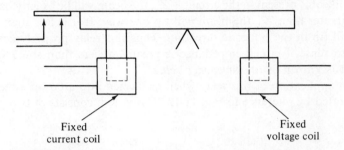

FIG. 11-10. A relay actuated by a current coil and restrained by a voltage coil.

$$K_i I^2 = K_v V^2 \qquad (11\text{-}9)$$

$$\frac{V^2}{I_2} = \frac{K_i}{K_v}$$

$$\frac{V}{I} = \sqrt{\frac{K_i}{K_v}} \qquad (11\text{-}10)$$

If this relay is connected into a circuit as shown in Fig. 11-11, the beam will balance if the ratio V/I is as given in Eq. (11-10). This means that for balance

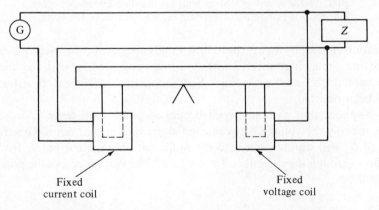

FIG. 11-11. A relay responsive to the value of the impedance in a a power circuit.

the impedance Z must have the value

$$Z_B = \sqrt{\frac{K_i}{K_v}}$$ (11-11)

If the value of Z present in the circuit is Z_B, the beam will be exactly balanced. If Z is greater than Z_B, the beam will tip one way. If it is less than Z_B, the beam will tip in the opposite direction. Thus this relay properly connected is able to sense if circuit impedance is greater or less than some specified value. It is known as an *impedance relay*.

An application of this relay when connected into a power system may be illustrated by reference to Fig. 11-12. The system consists of two generat-

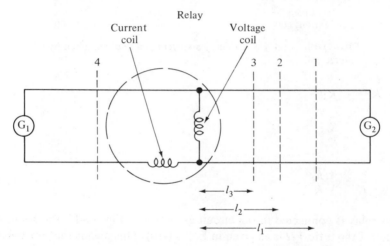

FIG. 11-12. A power system with an installed impedance relay. The relay is responsive, dependent on the location of the fault.

ing stations joined by an overhead line. Loads are not shown. Near the center of the line a relay is connected with its current coil in series with the line and its potential coil across the line. For simplicity, instrument transformers have been omitted.

Suppose that a short circuit occurs at point 1, distance l_1 from the relay, making the voltage between conductors zero at that point. Generators G_1 and G_2 will supply current to the fault, but only the current I_1 from G_1 will flow through the current coil of the relay. The voltage across the potential coil will be

$$V_P = I_1 Z_{l_1}$$ (11-12)

where Z_{l_1} is the impedance of the line from the relay to the fault point. Its value is

$$Z_{l_1} = l_1 z_u \qquad (11\text{-}13)$$

in which z_u is the impedance per unit length of line.

At some particular length of line, l_2,

$$l_2 z_u = Z_B = \sqrt{\frac{K_i}{K_v}} \qquad (11\text{-}14)$$

With the fault at position 2, the relay beam will just balance. If the fault is moved to position 3, the relay beam will tip to the left (Fig. 11-10) to close contacts. It is seen, therefore, that the relay will close contacts for any fault less than distance l_2 from the relay location. It will also respond to a fault at position 4, in which case only short-circuit current I_2 supplied by generator G_2 will pass through the relay. The impedance relay is influenced by the scalar value of the voltage and the scalar value of the current and is in no way influenced by the phase relation between these quantities.

The impedance of a line is made up of two parts, a real part R and an imaginary part jX (Fig. 11-13):

$$Z = R + jX \qquad (11\text{-}15)$$

However, the relay responds only to the scalar value of the impedance:

$$Z = \sqrt{R^2 + X^2} \qquad (11\text{-}16)$$

The operating characteristics of the impedance relay may be displayed on an

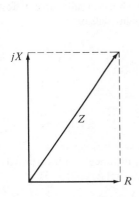

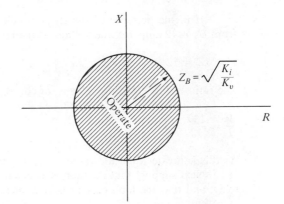

FIG. 11-13. An impedance diagram showing R, X, and Z.

FIG. 11-14. A diagram in the R-X plane showing the region of operation of an impedance relay.

R–X diagram as a circle of radius Z_B, centered at the origin, as shown in Fig. 11-14. At any time that the voltage on the potential coil of the relay and the current through the current coil of the relay are such as to indicate an impedance that falls within the circle, the beam of the relay will tip to the left and so close contacts.

EXAMPLE 11-1. Suppose that an impedance relay is designed with characteristics

$K_i = 128$

$K_v = 2$

Then

$$Z_B = \sqrt{\frac{128}{2}} = \sqrt{64} = 8$$

Let it be connected, as shown in Fig. 11-12, into a circuit in which the impedance per unit length is

$z = 0.2 + j0.75 = 0.77 \; \underline{/75°} \; \Omega/\text{mile}$

If a short circuit occurs at position 1, a distance

$l_1 = 15$ miles

the impedance of the line from the relay to the fault point is

$Z_{l1} = 0.77 \times 15 = 11.5 \, \Omega$

If in the presence of the short circuit, the short-circuit current flow from G_1 is 12 amperes, the voltage at the relay point will be

$V = IZ = 12 \times 11.5 = 139$ V

The ratio of the V/I at the relay is, of course,

$$\frac{V}{I} = \frac{139}{12} = 11.5$$

As this is greater than the value of Z_B, the relay will not operate.

Next suppose that the fault is at position 3, distance 4 miles from the relay, and that the fault current is 18 amperes. The relay voltage is

$V = Izl_3 = 18 \times 0.77 \times 4 = 55$ V

The impedance "seen by the relay" is

$$Z_{13} = \frac{V}{I} = \frac{55}{18} = 3.1 \ \Omega$$

As this value is less than Z_B, the relay will operate.

If the fault point is at a distance l_2 from the fault such that

$$Z = zl_2 = Z_B$$

then

$$l_2 = \frac{Z_B}{z} = \frac{8}{0.77} = 10.4 \text{ miles}$$

Now the relay beam will be just at the balance point.

It may be noted that the angle of the line impedance did not enter into the computations; only the *scalar* value was considered.

If the fault had been at position 4, the short-circuit current through the relay would have been supplied from generator G_2. The voltage on the relay, and its response, could have been calculated as in the previous cases. If for any combination of V and I, the ratio

$$\frac{V}{I} = Z_B$$

the relay will be on the balance point.

e. Voltage Coil and Watt Element. A balance beam relay may be constructed with a voltage coil and a watt element as indicated in Fig. 11-15. At the balance point, forces are in equilibrium, or

$$K_w V_P I \cos (\phi - \theta) = K_v V_v^2 \tag{11-17}$$

This type of relay might be connected as shown in Fig. 11-16. Now

$$V_P = V_v = V$$

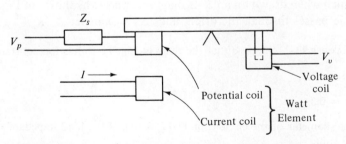

FIG. 11-15. A relay actuated by a watt-element and restraind by a voltage coil.

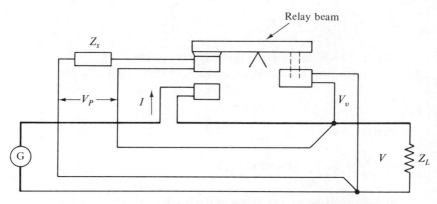

FIG. 11-16. A relay of the type shown in Fig. 11-15 connected to respond to the current and voltage of a power system. The circuit connection.

Solving Eq. (11-17) for the impedance seen by the relay under the condition of balance we find

$$\frac{V}{I} = Z_B = \frac{K_w}{K_v} \cos(\phi - \theta) \tag{11-18}$$

If Z_s is chosen such that the current in the potential coil is in phase with the voltage V_1, then

$$\theta = 0 \tag{11-19}$$

and

$$Z_B = \frac{K_w}{K_v} \cos \phi \tag{11-20}$$

It is seen that the magnitude of Z_B depends on ϕ, the angle between the current and the voltage associated with the load impedance Z_L. ϕ is, in fact, the angle of this impedance. Equation (11-20) is the polar form of a circle which when drawn on an R–X diagram appears as shown in Fig. 11-17. The circle passes through the origin and has a diameter K_w/K_v.

EXAMPLE 11-2. Suppose that the relay in Fig. 11-16 has the constants

$K_w = 50$

$K_v = 0.5$

The diameter of the circle (Fig. 11-17) is 100. If the load impedance Z_L has the value

$Z = 90 + j0$

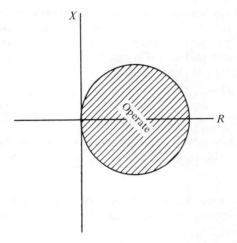

FIG. 11-17. The region of operation of the relay of Fig. 11-16 for $\theta = 0$.

its value falls within the circle and the relay will operate. On the other hand, if the load impedance Z_L is

$$Z = 0 + j20$$

its value falls outside the circle and the relay will *not* operate.

If Z_s is of such a value that θ is not zero, the circle shifts as shown in Fig. 11-18 with diameter unchanged and still passing through the origin.

If a relay of this type is used to detect short circuits in a line section, it may be desirable to adjust Z_s to such a value that

$$\theta = \tan^{-1}\frac{X_{\text{line}}}{R_{\text{line}}} \qquad\qquad (11-21)$$

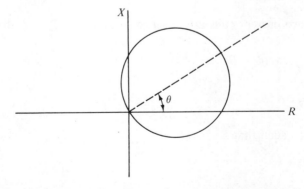

FIG. 11-18. The region of operation of the relay of Fig. 11-16 with θ non-zero.

Then the Z of the line, for any value of line length, will fall on the circle diameter, providing the situation with the most reliable response.

This relay might be connected into a system with two power sources in a fashion similar to that shown in Fig. 11-12. It may be seen that the relay responds for values of l such that

$$0 < lz < \frac{K_w}{K_v} \tag{11-22}$$

The relay does not respond if lz is greater than K_w/K_v, and it does not respond if l is negative. This relay is sometimes spoken of as a *directional distance relay*.

f. Current and Watt Coils. A balance-beam relay may be constructed with a current coil and a watt element as shown in Fig. 11-19. When forces are balanced,

$$K_i I_c^2 = K_w V I_w \cos(\phi - \theta) \tag{11-23}$$

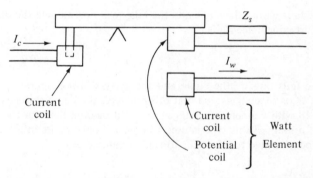

FIG. 11-19. A relay actuated by a current coil and restrained by a watt element.

If the two current coils are in series so that $I_c = I_w = I$, then

$$\frac{K_i}{K_w} = \frac{V}{I} \cos(\phi - \theta)$$

$$\frac{K_i}{K_w} = Z \cos(\phi - \theta) \tag{11-24}$$

If Z_s is chosen such that

$$\theta = 0 \tag{11-25}$$

$$\frac{K_i}{K_w} = Z \cos \phi = R \tag{11-26}$$

When a relay of this type is connected in a power circuit supplying an impedance Z_L, the balance point is determined by the resistance component of the impedance, the reactive component producing no influence. The performance characteristic may be shown on an R–X diagram (Fig. 11-20). From this it is seen that the relay will respond under all conditions in which current

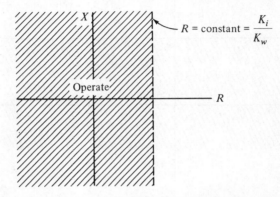

FIG. 11-20. The region of operation for a resistance distance relay.

and voltage values indicate a resistance of any value less than K_i/K_w, including negative values.

If this relay is adjusted to make

$$K_i = 10 \qquad K_w = 5$$

then $R = 2$.

Such a relay would close contacts in a circuit in which the voltage and the current were of such magnitudes and phase position as to make the impedance seen by the relay appear to be $1.2 + j20$. It would not close contacts if the impedance was $2.2 + j1.0$. A current and voltage condition that makes the impedance appear to be $-2.2 - j1.0$ would cause relay operation. This relay might be termed a *resistance distance relay* because it is responsive to particular values of resistance in the circuit to which it is connected.

If the impedance Z_s of the watt element is made highly inductive such that

$$\theta = 90° \tag{11-27}$$

$$\frac{K_i}{K_w} = Z \cos(\phi - 90)$$

$$\frac{K_i}{K_w} = Z \sin \phi = X \tag{11-28}$$

The response of this relay is as shown in Fig. 11-21. The relay will operate for all values of X less than K_i/K_w, including negative values. This is known as a *reactance distance relay*.

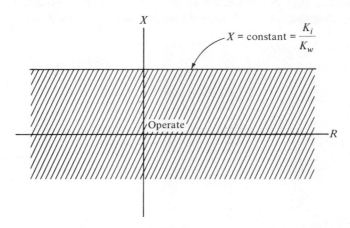

FIG. 11-21. The region of operation of a reactance distance relay.

Relays of this type have an advantage when used to measure the distance to the point of fault on a transmission line. The inductive reactance of a faulted line is directly proportional to line length. The resistance of the faulted circuit may include the resistance of the arc at the fault point and, in some instances, the resistance of a tower footing or other connections to the earth (see Chapter 13). Hence a distance measurement by a reactance relay is more reliable than a measurement made by a simple impedance relay. For example, a reactance relay would give the same response to an impedance of $Z_L = 2 + j12$ as to an impedance of $Z_L = 15 + j12$.

11-6. RELAY CONNECTIONS TO POWER EQUIPMENT

In Section 11-5 the relay current coils were shown connected in series with the power circuits while the potential coils were connected directly across the power-circuit terminals. These direct connections would be found only in low-voltage, low-power circuits. As indicated earlier, most instrument current coils are designed for nominal operation at 5 amperes; most potential coils are designed for nominal operation at 120 volts. If power-system values exceed these standard values, instrument transformers or similar devices must be employed. Figure 11-22 shows a balance-beam impedance relay connected to a 24-kV power circuit. A 300/5-ampere current transformer provides a current of appropriate value to the current coil of the relay. A 24,000/120-volt potential transformer provides voltage of proper value to

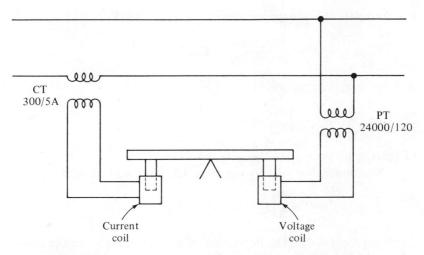

FIG. 11-22. An impedance relay connected to a power circuit through instrument transformers.

the potential coil of the instrument. For such an arrangement we may write

$$V_{line} = V_{relay}K_{pt} \qquad (11\text{-}29)$$

$$I_{line} = I_{relay}K_{ct} \qquad (11\text{-}30)$$

where K_{pt} and K_{ct} are the potential transformer ratio and the current transformer ratio, respectively. The apparent impedance of the line as indicated by voltage and current in the line is

$$Z_{line} = \frac{V_{line}}{I_{line}} = \frac{V_{relay}K_{pt}}{I_{reley}K_{ct}} \qquad (11\text{-}31)$$

Equation (11-32) shows the relations between the impedance of the line as indicated by its voltage and current, and the impedance seen by the relay as indicated by the voltage and current impressed across it.

$$Z_{line} = Z_{relay}\frac{K_{pt}}{K_{ct}} \qquad (11\text{-}32)$$

In the example shown in Fig. 11-22, a fault at some distance from the station might result in a current of 5000 amperes simultaneously with a voltage of 7500 volts. This would indicate a power-line impedance to the point of fault of

$$Z_L = \frac{V}{I} = \frac{7500}{5000} = 1.5\ \Omega$$

On the relay, the voltage would be

$$V_{relay} = \frac{V_{line}}{K_{pt}} = \frac{75,000}{24,000/120} = 37.5 \text{ V}$$

and the relay current would be

$$I_{relay} = \frac{I_{line}}{K_{ct}} = \frac{5000}{300/5} = 83.3 \text{ A}$$

(if there is no ratio error. See Fig. 10-7.)

The impedance seen by the relay due to conditions at its terminals is

$$Z_{relay} = \frac{V_{relay}}{I_{relay}} = \frac{37.5}{83.3} = 0.45 \; \Omega$$

Applying Eq. (11-32), the impedance of the line may be recalculated from relay values and instrument transformer ratios as

$$Z_L = 0.45 \frac{24,000/120}{300/5} = 1.5 \; \Omega$$

the correct value.

Other types of relays might be similarly connected to the power system through instrument transformers. The performance of the relays as determined by the electrical quantities supplied to them may be interpreted in terms of the performance of the power equipment by a suitable consideration of the instrument transformer ratios.

11-7. PROTECTION OF GENERATORS

This section presents for illustration several schemes for protecting generators by relays which are able to detect improper circuit behavior. Various other schemes are in use. The systems presented are somewhat oversimplified, for in each case instrument transformers are shown connected to the relays in the simplest arrangement possible. Actually several different types of relays may protect the same machine and one instrument transformer may supply several different relays as well as indicating instruments.

In previous discussions, current transformers were shown as two winding transformers (which they are), as indicated in Fig. 11-23a. For simplicity, henceforth current transformers will be indicated as shown in Fig. 11-23b. The diagram shows only the secondary of the transformer.

a. Generator Differential Protection. Current differential protection is a scheme by which the currents measured at two points

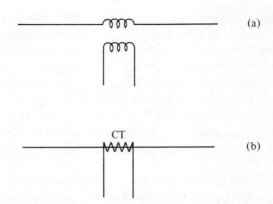

FIG. 11-23. Representation of a current transformer. (a) Two-winding representation. (b) Conventional representation.

are compared. These currents under normal conditions are equal but may become unequal on the occurrence of a fault. This scheme applied to one winding of a generator is shown in Fig. 11-24. A current transformer at the neutral end of the winding has its secondary connected in series with an identical transformer at the terminal end of the winding. A current relay is connected as shown. In normal operation the currents at both ends of the winding are equal and in phase. The secondary current around the loop then is the same at all points and no current flows through the relay coil.

If a fault develops at M (within the winding) which causes current to flow from the winding to the grounded frame or to another winding, the current at the neutral end differs from the current at the terminal end. The current in the secondary circuit to the left of the relay must correspond to the current in the neutral end of the winding, while the current to the right of the relay must correspond to the current at the terminal end. The difference

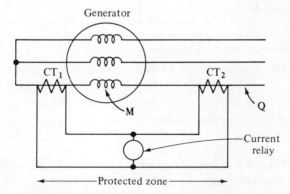

FIG. 11-24. Current differential protection of a generator. Protection on only one conductor is shown.

of these two currents must pass through the relay current coil. This relay then closes its contacts to clear the generator from the system. This usually involves opening the generator breaker, opening the field circuit, and closing the throttle on the prime mover driving the generator.

Note that this relay system responds to a ground fault at any point between CT_1 on the generator neutral and CT_2 on the machine terminal. The system does not respond to a fault at Q, for now the currents in CT_1 and CT_2 are identical and the difference current to the relay is zero. The region in which a fault causes operation of a relay is known as the *protected zone*.

The above arrangement assumes that the two current transformers CT_1 and CT_2 are identical, a situation that really never exists. If ratio errors in one are slightly different from those in the other, relay current may be present during normal operation and particularly during faults external to the generator. To avoid this difficulty, the relay is modified to respond to the differential current in terms of its fractional relation to the current flowing in the generator winding (Fig. 11-25). Thus under heavy load, a greater dif-

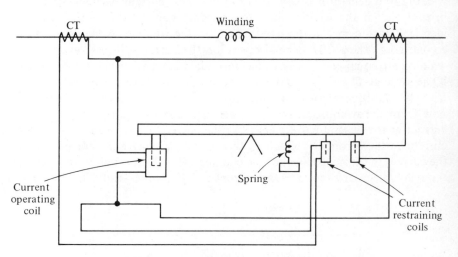

FIG. 11-25. A percent differential protection relay system.

ferential current through the relay operating coil is required for its operation than under light load conditions. Such a relay is termed a *percent differential relay*.

b. Split-Winding Protection. A balanced winding differential protection scheme may be used on large generators which have two parallel paths through the machine for each winding. As shown in Fig. 11-26, a current transformer connected in each branch of the winding permits a

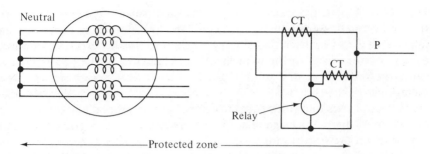

Neutral

CT

P

CT

Relay

Protected zone

FIG. 11-26. Protection of a generator with two paths for each winding.

comparison of the path currents. If these two path currents differ by a predetermined amount, the relay operates to clear the generator from the system. A percent differential relay may be used in this application.

The protected zone for the system of Fig. 11-26 extends from the machine neutral to the current transformers on the machine terminals. A fault at *P*, outside the protected zone, will not cause operation of the relay shown. An additional relay system would be necessary to initiate the clearing of a fault at this point.

c.   Ground Fault. Figure 11-27 shows a scheme for protecting a generator against a ground fault on one of its terminals (as shown by the

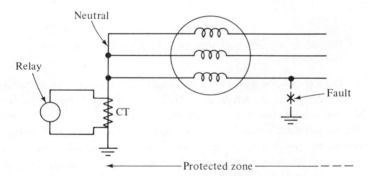

Neutral

Relay

CT

Fault

Protected zone

FIG. 11-27. Ground fault protection of a generator (neutral current measurement).

dashed line) or within the winding itself. Such a condition will cause current to flow in the neutral ground terminal, which will be detected by an overcurrent relay. The protected zone extends from the machine neutral to an indefinite distance out on the lines connecting directly to the machine terminals.

Another scheme for generator ground-fault protection is shown in

Fig. 11-28. A potential transformer is connected between the generator neutral and ground. Across the secondary of this transformer is connected an impedance Z and an overvoltage relay, as shown. Under normal conditions, a small current flows in the impedance as unbalances in the power system cause the machine neutral to be slightly different from zero potential. The voltage across the relay is low. On the occurrence of a ground fault, the machine neutral rises to a value equal to normal line-to-ground voltage, causing operation of the overvoltage relay. The relay operates for a ground fault at any point on the system directly connected to the generator terminals. This scheme of protection has the advantage that the current in the fault is low, being controlled in magnitude by the value of the impedance Z in the potential transformer secondary.

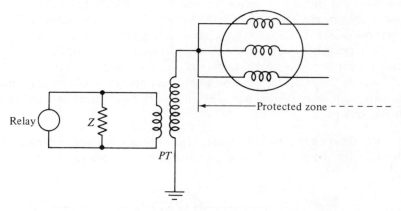

FIG. 11-28. Ground fault protection of a generator (neutral voltage measurement).

d. Ground-Fault Differential. Another scheme for generator-fault protection is shown in Fig. 11-29. The secondaries of each of the three current transformers connected at the terminals of the machine are connected in parallel. Under ordinary operation, the currents flowing in the generator leads and hence the currents flowing in the secondary windings add to zero and no current flows to the relay. Similarly, under normal conditions the neutral current is zero and no current is supplied by the neutral lead current transformer to the relay. If a ground fault develops at A external to the generator protective scheme, the sum of the currents at the terminals of the generator is exactly equal to the current in the neutral connection and the current supplied to the relay is zero. However, for a ground fault at B or within a winding, these currents are not equal and the difference current flows through the relay. Operation of the relay will clear the machine from the system. The protected zone is limited to the region between the neutral and the current transformers on the machine terminals.

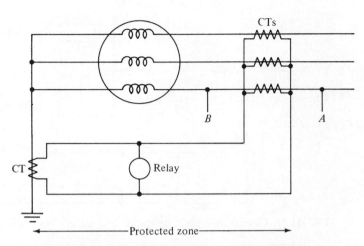

FIG. 11-29. Differential ground fault protection.

e. Other Relay Schemes. Other relay schemes used for generator protection include overcurrent protection, reversed power, reversed vars, excess temperature, loss of bearing oil pressure, and many other abnormal operating conditions.

11-8. BUS PROTECTION

Two methods of bus protection are described: one a differential scheme, the other a fault bus scheme. Various other arrangements are in use.

a. Differential Protection. In normal operation the sum of all currents coming to a bus must equal zero. If the sum of these currents (for a given conductor) is not equal to zero, it must be concluded that there is an unwanted connection (short circuit) to ground or to another one of the three conductors. Figure 11-30 shows the scheme of protection for a bus fed by a generator and supplying two lines. The system is shown for the A conductor. Similar systems would be required for the other two conductors. The secondaries of the current transformers in the generator lead, in line 1, and in line 2 are all connected in parallel. The protective relay is connected across this parallel connection. A current in the relay indicates a fault within the protected region and initiates opening of the generator breaker and each of the line breakers.

b. Fault Bus Protection. It is possible to design a station so that the faults which develop are entirely line-to-ground faults or at least the probability of such faults is much greater than the probability of line-to-

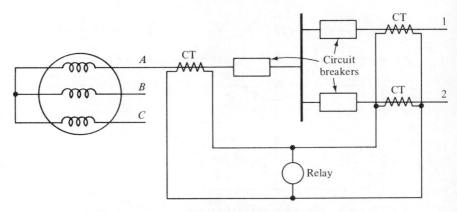

FIG. 11-30. Current differential protection of a station bus.

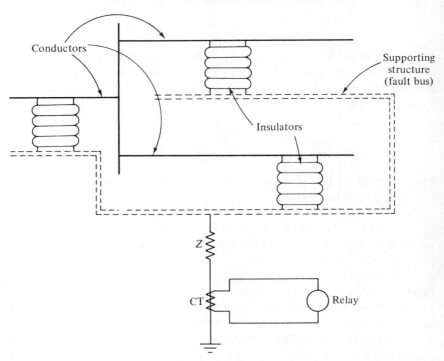

FIG. 11-31. Fault bus protection.

line faults. Particularly on generator buses and other high-capability, relatively low voltage circuits, a grounded metal barrier may surround each conductor throughout its entire run in the bus structure. With this arrangement every fault that might occur must involve a connection between a conductor and a grounded metal part. A somewhat less certain arrangement may be obtained in open construction by making spacing between adjacent conductors great compared to spacing over insulators to ground. Again faults that occur will almost certainly be from a conductor to the supporting structure.

In each of the preceding cases the metal supporting structure, known as the fault bus, is grounded through a current transformer as shown in Fig. 11-31. A fault involving a connection between a conductor and the grounded supporting structure will result in current flow through this current transformer and actuation of the relay connected to its secondary. Operation of this relay will result in tripping all breakers connecting equipment to the bus. An impedance Z in series with the current transformer serves to limit the value of the short-circuit current during a line-to-ground fault.

11-9. TRANSFORMER PROTECTION

Transformers are very commonly protected by current differential schemes such as the one illustrated in Fig. 11-32. Neglecting exciting currents,

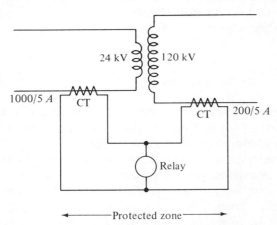

FIG. 11-32. Current differential protection of a transformer.

the current in the primary of the transformer is related to the current in the secondary by the turns ratio. This ratio must be considered in selecting the current transformers for the current differential protective scheme. As shown in the diagram, the current transformer on the 24-kV winding of the transformer has a rating of 1,000/5 amperes. On the 120-kV side of the transformer the current transformer has a rating of 200/5 amperes. During normal loading, the secondaries of these current transformers carry identical currents and may be interconnected with a current differential relay, as shown in Fig. 11-32. If a fault develops within the transformer, the currents will not balance and the differential relay functions to clear the breakers on both sides of the transformer. The protected zone extends from the current transformers on the high-voltage side to those on the low-voltage side of the power transformer.

Current differential relays may function improperly on closing of the breaker (either high side or low side) to energize the transformer after it has been out of service. Depending on the condition of the residual magnetism left in the transformer core due to previous operation and the point on the wave at which the closing breaker makes contact, initial magnetizing currents may be quite large, as much as eight times normal full-load current. These initial in-rush currents quickly diminish in value to the normal no-load excitation currents. However, since these initial in-rush currents exist in one winding only, an unbalance of currents between primary and secondary exists which may cause tripping of the differential relay. To avoid this improper tripping on closing, differential relays are sometimes made slow in operation for a few seconds following the closing of the breaker which reenergizes the transformer. In some systems, advantage is taken of the fact that initial in-rush currents are nonsinusoidal. Hence it is possible to design a relay system that discriminates between them and the nearly sinusoidal currents of a short circuit.

11-10. RELAY PROTECTION OF LINES

A great number of schemes are in use for the protection of overhead line and cable circuits. A few of these are discussed in this section. The descriptions presented pertain essentially to general principles rather than to details of the protection systems. In some cases two or more systems are used in combination to accomplish the desired selection. In other cases one complete system may be superimposed on top of another, both of which are capable of affording circuit protection. Duplicate, or redundant, protection systems are in use on many of the most important high-voltage circuits.

a. Inverse-Time Overcurrent Protection. As shown in Section 11-3b, a simple overcurrent relay may be modified to make its response time inversely proportional to the value of the current above minimum pickup. Such a relay has a performance characteristic curve, shown in Fig. 11-33. At values of current less than pickup, the relay never operates. At higher values the time of operation decreases steadily with increase of current.

It is important to note that the melting time of the current-sensitive element in a fuse may be described by a characteristic similar to that of Fig. 11-33. At currents less than the normal rated value, the fuse will not blow.

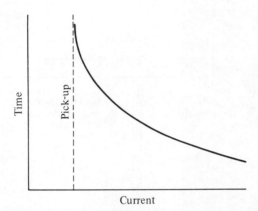

FIG. 11-33. The time-current characteristic of an inverse-current relay.

At currents above this value the current-sensitive element will melt, the time for melting being inversely related to the current value. For this reason, the present discussion of line protection by inverse-time overcurrent relays is equally applicable to circuit protection by fuses.

In applying relays for circuit protection, it is necessary to consider not only the time for the relay to operate but the time for the breaker to clear the fault after its trip coil is energized. In the case of the fuse, it is necessary to consider both the time to melt the fusible link and the time to clear the arc following the separation of contacts. These periods are shown diagrammatically in Fig. 11-34a and b. It is important to note that in the relay–breaker combination (Fig. 11-34a) the operation is irretrievably initiated when the relay contacts close, although it is not completed until the arc in the breaker is interrupted. Similarly, in the case of the fuse (Fig. 11-34b), the fuse operation is irretrievably initiated when the fusible link melts, but the operation is not completed until the arc between the opening contacts is extinguished.

The importance of the behavior of breakers (or fuses) is made evident by considering the protection of a small power system supplied by a single source and provided with several sectionalizing points (Fig. 11-35). Although there may be loads at numerous points, their presence is neglected, the assumption being that currents due to faults are much greater than the normal-load currents.

In Fig. 11-35, sectionalizing points are indicated at *A*, *B*, and *C*. A protection scheme is desired in which, on the occurrence of a fault, a minimum interruption to the system results. For example, with a fault at *P*, the breaker (or fuse) at *C* should open. However, if it fails to clear, the breaker at *B* should open as backup protection. To accomplish this behavior, the relays at the three different locations require different trip settings. If fuses

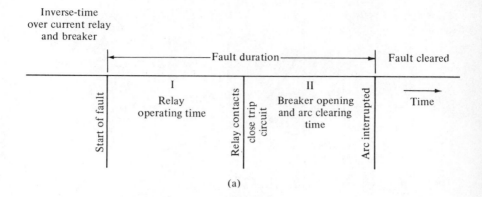

(a)

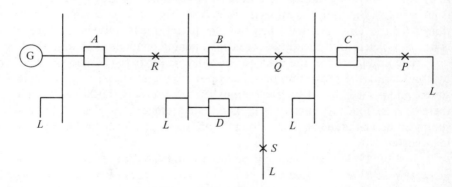

(b)

FIG. 11-34. Diagramatic representation of the sequence of events in clearing a fault. (a) Relay and breaker. (b) Fuse.

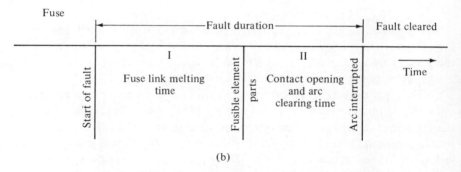

FIG. 11-35. A power system for describing relay selectivity and backup protection.

are used, the fuse characteristics must be similarly selected. The discussion from here on refers to a breaker-protection scheme, although all statements apply equally to a fuse-protection scheme.

Suppose the relays at points A, B, and C are given settings such that their performance characteristics are as indicated by (a), (b), and (c) of Fig. 11-36. The solid lines are the relay tripping times, whereas the dashed lines are obtained by adding the breaker clearing time, assuming it to be constant under all conditions.

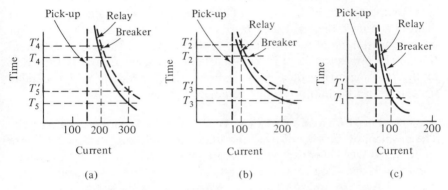

FIG. 11-36. The operating characteristics of the relay–breaker combinations of the system, Fig. 11-35.

Suppose that with a fault at P the short-circuit current is 100 amperes. The relay at C will close its contact in time T_1 and the breaker will clear the arc in time T_1'. With this same current the relay at B will close its contact at time T_2' and the breaker will clear the fault at time T_2'. The breaker at A would not operate with 100 amperes flowing in it, as this is less than the pickup current of the relay at A.

Note that the time T_1', the total fault-clearing time of the relay–breaker combination at C, is less than T_2, the relay time at B. If all equipment functions satisfactorily, the breaker at C operates to clear the fault and the breaker at B remains closed. However, if (due to a malfunction) the breaker at C fails to clear the fault, the breaker at B clears it in time T_2'. With this particular fault the breakers at B and C are properly coordinated. However, if T_1' had been greater than T_2, both breakers, B and C, would operate, interrupting unnecessarily any load connected between the two breakers.

A fault at Q results in a higher value of fault current, because the line impedance from the generator to the fault point is less than before. Assuming the fault current to be 200 amperes, the relay at B closes contacts in time T_3 and the breaker clears the fault at time T_3'. Similarly, with 200 amperes fault current the relay at A will close contacts in time T_4, and the breaker will clear the fault in time T_4'. Since T_3' is less than T_4, the proper functioning of the breaker at B results in the fault being cleared before the relay contacts at breaker A are closed. Hence breaker A normally will not operate.

A fault at S on the branch circuit results in behavior similar to that at Q, except that breaker D should respond instead of breaker B.

A fault at R, with a fault current assumed to be 300 amperes, will result in the relay at A closing contacts in time T_5, with breaker operation clearing the fault at time T'_5.

The above example illustrates a function known as *backup protection*, which is very important in relay-system design. Note that a fault at P combined with the failure of operation of the breaker at C results in a slightly delayed action of the breaker at B. If a fault at Q is combined with the improper functioning of the breaker at B, the trouble will be cleared by operation of the breaker at A. It is seen that faults are cleared even in the event of the malfunctioning of a relay or breaker. All properly designed relay systems afford backup protection to adjacent zones.

b. Impedance Relays With Directional Elements.

The problem of relaying of overhead lines and cable circuits becomes more complicated when there are two or more generating stations on the system. Figure 11-37 shows a two-generator system with several sectionalizing points.

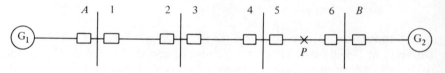

FIG. 11-37. A power system with 2 sources and 4 sectionalizing points.

The loads at each station are not shown. Impedance relays with inverse time delay and a directional element are assumed at each breaker. Both must close contacts to trip the breaker. It is assumed that the directional elements on all odd-numbered breakers will pick up on the occurrence of a fault to their right. The directional element on all even-numbered breakers will pick up for faults to their left. The selection of the eligible breaker that should trip in the event of a fault is determined by the impedance relays.

Consider a fault at P. The directional element of breaker 1 operates as the fault is to the right. The current and voltage conditions on the impedance relay are such as to indicate the impedance $z_u l_{1P}$, which is the impedance of the line from breaker 1 to the fault point P. Let it be assumed that this is outside the operating range of the impedance relay on breaker 1.

As viewed by breaker 2, the fault is in the wrong direction for tripping and the directional element does not operate, thereby blocking breaker trip.

At breaker 3 the directional element operates to permit tripping. The impedance relay responds to the impedance $z_u I_{3P}$. Assume that this is just barely within the operating range of the impedance relay. This relay tends to close its contacts but will be slow in operation.

At breaker 4, the fault is in the wrong direction for the directional element to close contacts.

At breaker 5, the directional element operates to permit tripping. The impedance seen by the impedance relay is $z_u I_{5P}$. As this impedance is low, it falls well within the region of operation of the impedance relay. Hence this relay operates quickly to provide trip current for the breaker. In case breaker 5 fails to trip, breaker 3 will trip after a time delay, thus providing backup protection.

At breaker 6 the directional element operates to permit tripping. The impedance relay on this breaker sees low fault impedance and quickly closes tripping contacts. Should this breaker fail to open, the protective scheme on generator G_2 should cause breaker B to open.

The scheme just described was illustrated by a relatively simple power system consisting of several series-connected line sections. The method is equally applicable to networks in which there may be multiple parallel lines, closed loops, and numerous generating stations. Each breaker is set to trip if its directional element responds, indicating the location of the fault to be in the appropriate direction. The impedance relay closes contacts quickly if current and voltage conditions indicate low impedance between the breaker and the fault. Tripping occurs after an appreciable time delay if the impedance indicates that the fault is beyond the next station.

The above description referred to the use of impedance relays, a type which has found very extensive application. Reactance relays, described in Section 11-5f, might have been used instead. The reactance relays have the advantage that they respond only to the reactance of the circuit, a value that is unaffected by the presence of fault circuit resistance, a factor which may be of importance if long arcs form the short circuit at the fault point.

It has been assumed that the impedance (or reactance)-measuring relays have inverse-time characteristics. Distance-measuring relays, in which the timing is by steps, have been designed and are in extensive use. If the impedance seen by the relay is in a certain low range, the closing of tripping contacts will be without intentional time delay, or instantaneously. If the impedance measured is in a somewhat greater range, a fixed time delay is introduced. If the impedance is in a still greater range, a longer fixed time delay is introduced. If the impedance is beyond this third range, no operation results. Relays of this type are spoken of as *step-distance relays*.

In application the step-distance relay is frequently set to be instantaneous over about 80 per cent of the line it protects. If the measured impedance indicates the fault to be in the far end of the protected section or in the next section, tripping time is increased. The third step operation extends from the second circuit section into the third. This range is illustrated in Fig. 11-38. With this arrangement each breaker provides backup protection for two other breakers.

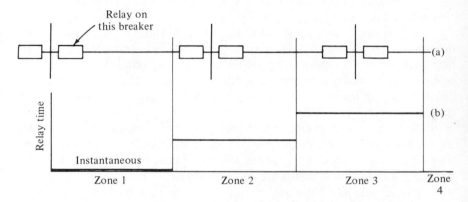

FIG. 11-38. Step-distance relays applied to a power system. (a) The system. (b) The time steps of one relay for faults at different locations.

c. **Directional Comparison.** The *directional comparison method* of relay protection is illustrated by means of Fig. 11-39. Here it is assumed that a fault exists at point *P*. The operation of the directional relay

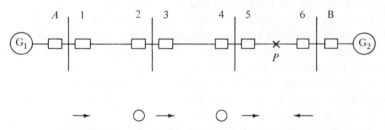

FIG. 11-39. A power system for illustrating a system of directional comparison protection.

element provided on each breaker is indicated by the arrow. A directional element that does not operate for a fault at *P* is indicated by the letter O. It is evident that if one could stand in the middle of a line section and view the directional elements on each end of the section, it would be possible to determine if the fault were in that section or external to it. The directional element of breaker 1 has operated. The directional element on breaker 2 at the other end of the line section is open, indicating that the fault is not in this section. Similarly, the closed position of the directional element on breaker 3 and the open position on breaker 4 indicates that the fault is external to the line between 3 and 4. The directional element on breaker 5 is closed, and similarly the directional element on breaker 6 is closed, indicating that the fault is between the two breakers and that both breakers should trip.

The basic principle of directional comparison is very simple and positive. A difficulty arises in attempting to view the directional elements at both ends of the lines simultaneously. If the line terminals are not far apart geographically, a wire communication circuit between them may permit automatic comparison of relay positions with high-speed removal of the faulted line section. However, if the line sections are 100 miles or more in length, wire communication channels may be expensive and perhaps unreliable. Here it may be desirable to utilize a channel in a high-frequency carrier system, which is transmitted over the conductors of the power circuit itself. Abnormal conditions existing on the occurrence of a fault activate relays on each of the breakers in line sections near the fault. These relays, unless blocked from operation, cause tripping of the breakers. The blocking signal is controlled by the directional relays on each breaker, and is transmitted from one end of a protected section to the other by carrier or by microwave. If a directional element determines that the fault is not in the direction of the protected section, a signal is transmitted which blocks breaker operation at both ends of the section. If the directional elements at both ends of the line indicate that the fault is in the direction of the protected section, no blocking signal is transmitted from either end, and both breakers trip. The sequence of events is made clear by study of Fig. 11-39.

At breaker 1 the directional element indicates that the fault may be in the line section from 1 to 2. This breaker trips if no blocking signal is received. No blocking signal is transmitted to breaker 2.

At breaker 2 the directional element shows that the fault is not in the line section. A carrier signal is transmitted which blocks tripping of both breaker 1 and breaker 2. Behavior at breakers 3 and 4 is similar to that at 1 and 2 respectively.

At breaker 5 the directional element indicates the possibility of a fault in the line section. No blocking signal is transmitted. The directional element at breaker 6 also indicates the possibility of the fault in the protected section. No blocking signal is transmitted. After a very short time delay, both breakers 5 and 6 trip.

The transmission of a *blocking signal* between stations (rather than a tripping signal) provides more reliable fail-safe operation. A fault in the middle of the line might interfere with carrier frequency transmission and so make impossible the transmission of a tripping signal. A blocking signal needs to be transmitted only if the line is in good condition.

A failure of the blocking signal equipment causes the opening of a line section on which no fault exists. However, this type of false operation is preferable to the failure to clear a faulted section.

d. Phase-Comparison Protection. The presence of a fault in a line section may be detected by comparing the phase position of the currents as observed at each of the two ends of the line, in a manner somewhat similar to differential protection. The comparison of these currents is made over carrier channels or microwave channels. The method is illustrated by reference to a short section of power system (Fig. 11-40). In column (a) behavior is described for a fault assumed at point P. The current flowing *into* the line section at circuit breaker A is as shown on line 1. If this current is of sufficient magnitude, a carrier signal is transmitted which goes to the relaying system of breaker A and over a communication circuit to the relaying system of breaker B. As indicated on line 3, carrier transmission is only during the positive half-cycle of the current wave.

The current flowing *into* the line section at breaker B is as indicated in line 2. This current is of different magnitude than the current at A but is almost in phase with it. If it is of sufficient magnitude, there will be carrier transmission to the relay equipment at B and over the communication circuit to the relay equipment at A. This transmission occurs only during the positive half-cycle of the current flow at B, as shown on line 4. The carrier transmissions at A from A and from B are added and rectified to yield an output signal, as shown on line 5. When the signal has a zero value for a specified time (measured in milliseconds) as at T, auxiliary equipment causes the trip circuit of the breaker to be energized. At breaker B, both signals are received and are processed in the same fashion.

Next consider a fault assumed at Q. In column (b), Fig. 11-40, behavior is described for this condition. The current flowing *into* the line section at breaker A is as shown in line 1. This causes carrier transmission as shown in line 3. This carrier transmission is received by the relay equipment at both breakers A and B.

At breaker B the current *into* the protected section is as indicated by line 2. Note that it is 180° out of phase with the current into the line section at A. This causes carrier transmission (during the positive half-cycle only), as shown in line 4. This carrier is transmitted to the relaying equipment at both A and B, as before. The sum of the A carrier and B carrier added and rectified results in a signal, as shown in line 5. This signal maintains a nonzero value throughout the entire cycle, preventing the auxiliary equipment from energizing the trip circuit to the breaker. Similar behavior at breaker B results in no signal to the trip-coil circuit of that breaker.

It may be noted that with this phase-comparison arrangement, a trip signal is available with a fault in the *line* section, but none is available if the fault is exterior. In the event of failure of the carrier equipment, the affected

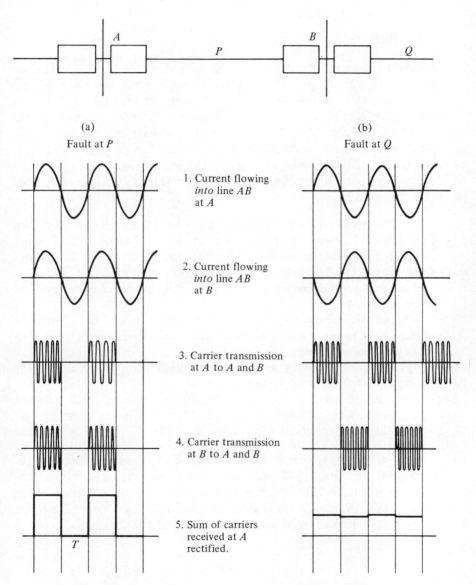

FIG. 11-40. A section of a power system for illustrating phase comparison protection.

line section will be opened even though no fault exists on it. Solid-state equipment applied to such a system provides very high speed reliable relay protection.

11-11. SOLID-STATE RELAYING

Designers of electronic circuits have developed systems using solid-state components that duplicate the functions of the many electromechanical relay types that have been discussed. The circuitry used for a particular function may be considerably different as developed by different designers. No attempt is made here to present a group of solid-state relay designs corresponding to the electromechanical ones described. For purposes of illustration, the broad operations necessary in the phase-comparison system are outlined. A few solid-state circuits providing some important relay functions are also presented.

a. Functions Necessary in the Phase-Comparison Method of Circuit Protection. A *level response circuit* recognizes that the current magnitude is indicative of a fault condition. The output of this level circuit is necessary to *permit* carrier transmission.

A *rectifier* produces an output only during the positive half-cycle of the current wave.

A *switching circuit* responds to signals from the level circuit and the rectifier circuit to cause transmission of carrier only during the positive half-cycle of the current wave.

A circuit is necessary to add carrier transmission received from the two ends of the protected section.

The added carriers must be rectified to produce either a constant output for blocking or a pulsating output allowing tripping depending on the time displacement of the two carriers.

b. Examples of Static Relay Equipment. A few simple circuits will illustrate typical relay equipment using solid-state devices. The following material is taken directly from a General Electric Co. publication, *Printed Circuit Cards for Static Relay Equipment* (GEK-7364). (The figure numbers have been changed to agree with the text.)

OPERATING PRINCIPLES

Although there are many varieties of printed circuit cards used in General Electric static relays, the following descriptions of typical logic, discriminator, and timing circuits should be sufficient to familiarize the user with the basic operating principles used.

LOGIC CARD

The logic of Figs. 11-41 and 11-42 is well suited for use as an example since it includes all three basic forms of logic; the AND, OR, and NOT functions. The (b) section of Fig. 11-41 illustrates the functional block representation of the three functions, while the (c) section illustrates the representation usually used on logic unit internal connection diagrams and overall logic diagrams for General Electric static relays. A mechanical contact equivalent is also shown in section (a) for comparative purposes.

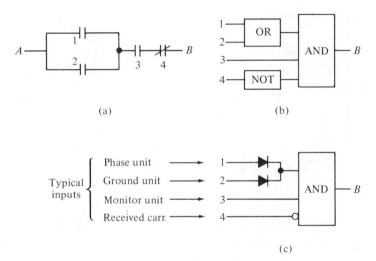

FIG. 11-41. Typical logic function.

In the logic diagrams, inputs 1 OR 2 AND 3 must be present, AND input 4 must NOT be present if there is to be an output at B. In the mechanical contact equivalent, relay 1 OR 2 AND relay 3 must operate, AND relay 4 must NOT operate if there is to be a circuit from A to B.

The circuit of Fig. 11-42 will produce the logic of the diagrams in Fig. 11-41. In this circuit, the output transistor Q_4 must be turned on to produce an output at B. Since Q_4 is a *PNP*, transistors Q_1 and Q_2 must be turned on if there is to be an output at B. Note that Q_1 and Q_2 are *NPN* transistors. This means that their bases must be positive in relation to their emitters which are biased slightly positive. Consequently, positive signals are required at 1 OR 2 AND 3 to turn on the transistors and cause current to flow in resistor R.

Note that the circuit is arranged so that Q_3 will short-down the input to Q_2 when there is a positive input at 4. This, of course, will prevent Q_2 from the turning on, even though there is an input signal at 3. The total circuit behavior, like that of Fig. 11-41, can be expressed by the statement: Inputs at 1 OR 2 AND 3 and NOT 4 are required to produce an output at B.

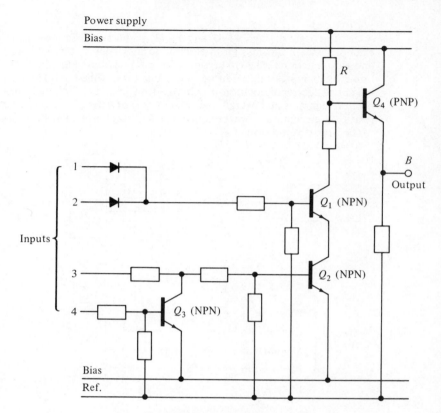

FIG. 11-42. Typical logic function.

The inputs indicated in Fig. 11-41c will be recognized as typical of the inputs to the comparer logic of a directional comparison carrier scheme.

In typical logic circuits, such as that previously described, the transistors are being used as switching devices. Hence any variation in amplification factors between units, or variations in magnitude of input signals, have a negligible effect on the overall operation of the circuit.

DISCRIMINATOR

The operation of a typical discriminator or level detector card is illustrated in Fig. 11-43. Here, response to a signal magnitude is involved so the signal to be measured must be compared to a regulated bias voltage. In a typical application, the quantity to be measured, the current I, is fed through the primary of a transactor T_1, which produces a secondary voltage proportional to I. A portion of this voltage, as determined by potentiometer P_1, is then coupled to the input pins 3 and 5 of the detector card through center-tapped transformer T_2.

In the standby condition, transistors Q_1 and Q_2 will be off and Q_3

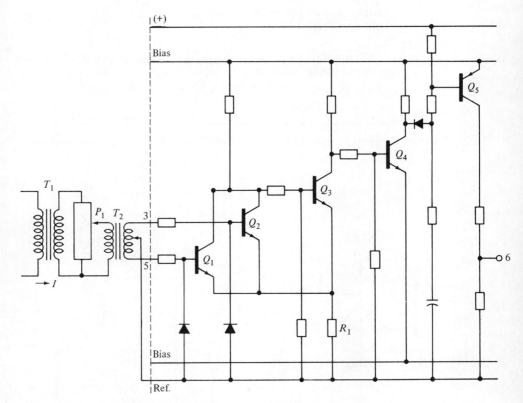

FIG. 11-43. Typical discriminator or level detector card.

will be on, because the base of this *NPN* transistor will be positive with reference to its emitter. With Q_3 conducting, the next *NPN* transistor, Q_4, will be off since its base will be negative with reference to its emitter. Consequently, the *PNP* transistor Q_5 will also be off, since its base will be positive with reference to its emitter, and there will be no output at point 6. Note that the current passed through resistor R_1 by Q_3 establishes the plus bias on the emitters of Q_1 and Q_2 and hence determines the input signal required at their bases for operation.

When the signal at point 3 (or 5) exceeds the bias on the Q_2 (or Q_1) emitter, the transistor will turn on and cut off Q_3. Note that as soon as Q_3 cuts off, the bias will decrease, resulting in a "snap action" of the detector. With Q_3 off, transistors Q_4 and Q_5 will turn on, producing an output at point 6. Note also that when Q_4 turns on, the capacitor will discharge.

When the sine-wave input to point 3 falls below the reduced bias on Q_2, that transistor will turn on again. However, the charging current into the capacitor will hold Q_5 turned on for 9 milliseconds. This overrides the interval between adjacent half-cycles so that the output will be continuous if each adjacent half-cycle is above the turn-on point of Q_1 or Q_2.

TIMER

A typical time-delay card having both operating and reset delays is shown in Fig. 11-44. This card uses an RC timing circuit (consisting of R_1 and C_1) to determine the operating time. An input signal at

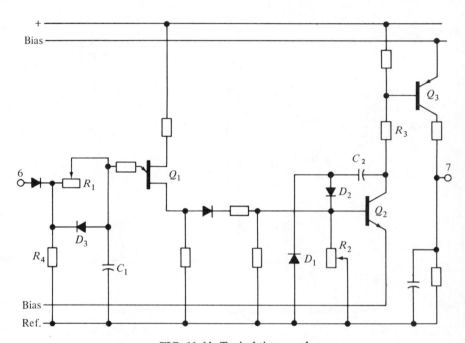

FIG. 11-44. Typical timer card.

point 6 will charge the capacitor through R_1. At the end of the delay interval, as determined by the R_1 setting, the unijunction transistor Q_1 will be turned on. This will cause the base on *NPN* transistor Q_2 to be positive with respect to its emitter. When Q_2 conducts, transistor Q_3 will also turn on, resulting in an output at point 7. Note that diode D_3 and resistor R_4 provide a low-resistance discharge path for capacitor C_1, resulting in fast reset if the input signal is removed before the end of the set time delay.

Once an output signal has appeared, time delay on reset is provided by capacitor C_2. In standby, C_2 is normally charged, but when Q_2 conducts, C_2 will discharge through Q_2, D_1, and the supply source. When the input is removed from point 6 causing Q_1 to cut off, capacitor C_2 will recharge through R_3, D_2, and adjustable resistance R_2. The charging current will hold Q_2, and hence Q_3, in the turned-on condition. The reset time (that is, the time until Q_2 cuts off) is determined by the setting of R_2.

PROBLEMS

11-1. The current in a certain ac circuit should be held at 800 A. Show a relay system which will sound an alarm if the current falls to 750 A. Choose a balance-beam relay showing the necessary coil or coils, spring, contacts, and any instrument transformers necessary.

11-2. Show the complete setup of a relay system for use in a 240-V ac circuit which will sound an alarm if the voltage rises above 260 V or falls below 220 V.

11-3. A gasoline-driven 120-V 5-kW ac generator is connected through a breaker to a large 120-V system. Show a relay system that will energize the trip coil of the breaker in the event that the gasoline engine runs out of fuel and power flows from the large systems to the 5-kW generator, driving it as a motor.

11-4. Show a relay system which will sound an alarm if the voltage of a circuit exceeds 70 V. The alarm should continue even though the voltage later drops below 70 V.

11-5. Refer to Fig. 11-5. What would be the result if, on closing the circuit breaker, the auxiliary contacts failed to close? Explain.

11-6. Refer to Fig. 11-5. Assume that with the circuit breaker closed, a test man, by mistake, opens the potential transformer secondary circuit. Explain the result.

11-7. Show a relay arrangement which might be used to sound an alarm 5 min after the power supply fails.

11-8. Refer to the circuit of Fig. 11-11. Tests on the relay itself show that the torque on the beam due to 5 A in the current coil is just balanced by that due to 120 V on the voltage coil. Which way will the beam tip if the load impedance Z is

a. $6 + j9$?
b. $15 + j20$?
c. $20 + j20$?

For this system what is the ratio K_i/K_v?

11-9. Refer to the system of Fig. 11-22, but assume the following values: PT ratio: 69,000/120, CT ratio: 400/5, $K_i/K_v = 36$. Assume that the voltage between conductors is 58 kV. What conductor current is necessary to produce a balance of the relay beam?

11-10. Draw a diagram showing the relay of Fig. 11-16 connected for operation on a 12-kV circuit on which the current is normally 200 A. Assume that the relay itself has 120-V-potential coils and a 5-A current coil. Assume that $\theta = 0$. What must be the value of the ratio K_w/K_v to make the relay responsive to line impedances from $Z = 0$ to $Z = 20$?

11-11. Refer to Fig. 11-15. The potential coil of the watt element has a resistance of 45 Ω and an inductance of 0.15 mH.

    a. What must be the value of the impedance Z_s to make $\theta = 0$?

    b. What must be the value of Z_s to make the relay match a line in which the impedance per mile is $z = 0.15 + j.75\ \Omega$?

11-12. A reactance relay is so constructed that K_i/K_w is 75. On test, 60 V is applied to the potential circuit while a current of 3.5 A lagging the voltage by 40° is passed through the current coils. Will the relay operate? Explain.

11-13. Draw a diagram of a relay system that will protect a generator, providing trip current to the breaker in case of

    a. a fault within the generator.

    b. excess load current.

    c. power flow from the system to the generator (reverse power).

Show all necessary CT's and PT's. Design the system on a line-to-neutral basis.

11-14. Refer to Figs. 11-35 and 11-36. Assume that with a fault at P, the relay contacts at breaker C close but the breaker is sluggish in operating. How much delay could be tolerated without resulting in the unnecessary opening of breaker B? If both breaker C and breaker B are sluggish, how much delay could be tolerated without the unnecessary opening of breaker A?

11-15. Refer to Fig. 11-39. Suppose that, owing to a failure of equipment, no blocking signal is transmitted from breaker 4. Explain the results.

chapter 12

Electrical
Insulation

12-1. PURPOSE

Insulation is required to keep electrical conductors separated from each other and from other nearby objects. Ideally, insulation should be totally nonconducting, for then currents are totally restricted to the intended conductors. However, insulation does conduct some current and so must be regarded as a material of very high resistivity. In many applications, the current flow due to conduction through the insulation is so small that it may be entirely neglected. In some instances the conduction currents, measured by very sensitive instruments, serve as a test to determine the suitability of the insulation for use in service.

Although insulating materials are very stable under ordinary circumstances, they may change radically in characteristics under extreme conditions of voltage stress or temperature or under the action of certain chemicals. Such changes may, in local regions, result in the insulating material becoming highly conductive. Unwanted current flow brings about intense heating and the rapid destruction of the insulating material. These insulation failures account for a high percentage of the equipment troubles on electric power systems. The selection of proper materials, the choice of proper shapes and dimensions, and the control of destructive agencies are some of the problems of the insulation-system designer.

12-2. INSULATING MATERIALS

Many different materials are used as insulation on electric-power systems. The choice of material is dictated by the requirements of the particular application and by cost. In residences, the conductors used in branch circuits and in the cords to appliances may be insulated with rubber or plastics of several different kinds. Such materials can withstand necessary bending, are relatively stable in characteristics, and are inexpensive. They are subjected to relatively low electrical stress.

High-voltage cables are subjected to extreme voltage stress; in some cases several hundred kilovolts are impressed across a few centimeters of insulation. They must be manufactured in long sections, and must be sufficiently flexible as to permit pulling into ducts of small cross section. The insulation may be oil-impregnated paper, varnished cambric, or synthetic materials such as polyethylene.

The coils of generators and motors may be insulated with tapes of various kinds. Some of these are made of thin sheets of mica held together by a binder, and others are of fiber glass impregnated with insulating varnish. This insulation must be capable of withstanding quite high operating temperatures, extreme mechanical forces, and vibration.

The insulation on power-transformer windings is commonly paper tape and pressboard operated under oil. The oil saturates the paper, greatly increasing its insulation strength, and, by circulating through ducts, serves as an agent for carrying away the heat generated due to I^2R losses and core losses in the transformer. The transformer insulation is subjected to high electric stress and to large mechanical forces. The shape and arrangement of conducting metal parts is of particular concern in transformer design.

Overhead lines are supported on porcelain insulators. Between the supports air serves as insulation. Porcelain is chosen because of its resistance to deterioration when exposed to the weather, its high dielectric strength, and its ability to wash clean in rain.

12-3. INSULATION BEHAVIOR

When insulation is placed between two metallic conductors A and B connected to a voltage source (Fig. 12-1), several phenomena associated with the insulation may be identified. The insulation, or dielectric, influences the capacitance between the plates, a current of low magnitude flows through the body of the insulation, a leakage current flows over the surface of the insulation, and if the voltage is great enough, sudden changes in the body of the insulation may make it highly conductive.

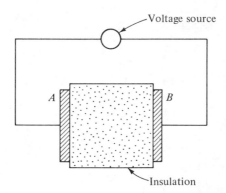

FIG. 12-1. Insulation under electrical stress between two metal electrodes.

a. Capacitance and Dielectric Hysteresis. As is well known, the presence of a dielectric between two conducting plates increases the measured capacitance between these plates. The behavior of the electrons and protons comprising the dielectric accounts for this capacitance increase. This phenomenon is worth investigating, for it explains some other characteristics of insulation of interest.

As discussed in Chapter 9, all matter is made up of protons, electrons, and neutrons. In the normal state, these particles are grouped as atoms or molecules in which the number of electrons (each with unit negative charge) and the number of protons (each with unit positive charge) are equal, therefore, each group is electrically neutral. However, each of these particles experiences a force due to its interaction with any charges placed on nearby plates. We say that the charged particles respond to the electric field set up between the plates.

In a perfect insulator, the electrons and protons are held together in the atoms and molecules and are not free to drift from one plate to another. However, in the presence of an electric field, they may move very slight distances, the electrons toward the anode, the protons toward the cathode. This situation is illustrated in greatly simplified form in Fig. 12-2a. This diagram shows a group of polar molecules. Each of these is neutral, but each has an extra electron at one end and an extra proton at the other. These may be moving with respect to each other but at some instant have a position as shown. Let us confine our attention to the polar molecule at P.

Next let charges be put on A and B by connection to a voltage source as shown in Fig. 12-2b. The molecule P rotates, taking up a new position with the electron displaced toward A and the proton toward B, under influence of the electric field.

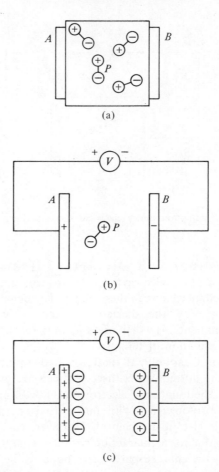

FIG. 12-2. The effect of electrical stress of insulation. (a) Some polar molecules under normal conditions. (b) The polar molecule *P* changes position under stress. (c) The charge layers adjacent to the electrodes.

The effect of the many, many electron–proton pairs in the dielectric volume, moving as shown by the example *P*, is to produce an effect shown in Fig. 12-2c. Adjacent to the anode *A* there are an excess number of electrons in the dielectric; near the cathode *B* are an excess number of protons in the dielectric. These charges partially neutralize the effect of the charge originally placed on the plate and additional charges move from the voltage source to *A* and *B* as the charges in the dielectric take on their new positions. Hence, for the same voltage between the plates, the charges that have moved from the source to the plates have been increased as a result of the presence of the dielectric. The capacitance between the plates is greater, therefore, than it would be without the dielectric.

Whenever a system of particles is moved from one position to another (such as system P from the position shown in Fig. 12-2a to that of Fig. 12-2b), there are forces which restrict the motion, and time is required to make the change. Such is the case in dielectrics. In some materials the change is made in a fraction of a microsecond; in others it may take several hours. During the period of change, the capacitance appears to increase and current flows in the external circuit. It is sometimes stated that charge is "soaking" into the dielectric. The phenomenon is known as dielectric absorption.

When the voltage source is disconnected from plates A and B and the voltage between them is made zero by a short-circuiting connection (Fig. 12-3a), the displaced particles in the dielectric tend to go back to their

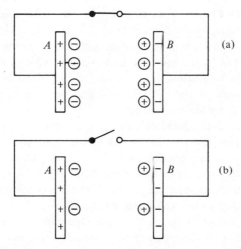

FIG. 12-3. Charge conditions in a capacitor that has been energized. (a) Immediately after being short-circuited. (b) After the short circuit is removed.

normal state. However, if it took a long time to get them oriented, it will take a long time to get them back into the normal state. Hence the condition shown in Fig. 12-3a may persist for some time, a charge remaining on each plate equal in effect to the charge remaining in the dielectric adjacent to the plates.

Next suppose that the short circuit is removed. Forces continue to restore the dielectric to its neutral state. With the circuit open, the charges held on the plates cannot be removed. As the dielectric returns to its normal condition, the trapped charges on the plates produce a voltage between A and B. This voltage may be a serious hazard to a workman who expected the capacitance between the plates to be discharged by the short-time application of a short circuit. This hazard is particularly serious on equipment of high capacitance such as high-voltage cables, static capacitors, and generator windings. For this reason, it is always desirable to keep such equipment continuously short-circuited when workmen are to be in physical contact with the presumably deenergized equipment.

Referring again to Fig. 12-2a and b, the movement of particles, such as the polar molecule *P*, may result in the movement of other nonpolar molecules. If the molecular motion is increased, the temperature of the material is increased. If the power supply is an ac source, each reversal of voltage will tend to cause a reversal of the position of the polar molecules and electrical energy from the source will be converted to heat in the insulation. This loss is known as *dielectric hysteresis*. It increases with frequency and with applied voltage. It must be considered in high-voltage cable design.

b. Conduction Currents. When voltage is applied between two plates separated by a dielectric (Fig. 12-1), those few free electrons that are present in the insulation drift from cathode to anode. This is termed a conduction current (from anode to cathode) and represents power loss into the insulation. In insulation, the number of free electrons is low, and as a result the resistivity of the material is high. The number of free electrons may be increased by several processes.

An increase in temperature causes an increase in the thermal agitation of the molecules of the insulation. As temperature increases, an increased number of electrons are broken loose from their parent molecules and are free to drift in the presence of an electric field. The resistivity of the material decreases with an increase in temperature. This is in contrast to metals, where resistivity increases with increase of temperature.

Insulating materials have chemical structures designed to have high electrical resistance. Changes in these structures may result in a substantial reduction of the resistance. Chemical changes may result from periods of elevated temperature, from the action of light, or from attack by certain chemicals.

The exposure of insulation to radiation particles (as from nuclear reactors) increases the number of free electrons, and so causes a temporary reduction of resistivity. Under extreme exposure, chemical changes resulting from radiation damage may permanently lower insulation resistance.

c. Surface Leakage Currents. Leakage currents flow along paths between electrodes over the surface of the insulating material. The magnitude of these currents is in no way related to the resistivity of the material itself. The value of the leakage current depends on the applied voltage, the insulation material, the surface contamination, and the moisture content of the air. On seriously contaminated high-voltage line insulator surfaces, leakage currents may be as much as 100 milliamperes.

d. Insulation Breakdown. Insulation may undergo a very sudden change in characteristics in a process known as *breakdown*. Consider the arrangement shown in Fig. 12-4. Two parallel-plane electrodes A and B

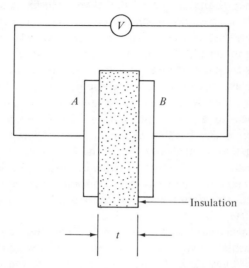

FIG. 12-4. Insulation of thickness t being stressed by applied voltage V.

are separated by a sheet of dielectric of thickness t. A variable voltage source V provides a difference of potential between A and B. Suppose the voltage is slowly raised. At first the conduction current is very low, perhaps measurable in microamperes. With increased applied voltage, the current suddenly increases, and the insulation takes on the character of a metallic conductor. This is termed insulation breakdown. On examination, a small damaged place may be found extending through the insulation sheet. Perhaps there will be some charring and perhaps there will be a hole.

The voltage at which such breakdown occurs is called the *breakdown*, or *puncture*, *voltage*, V_p, and the electric field intensity $\mathscr{E}_p$ at that point is known as the *breakdown gradient* or puncture strength of the insulation,

$$\mathscr{E}_p = \frac{V_p}{t} \tag{12-1}$$

where t is the insulation thickness.

The puncture strength of a particular sample is not a constant but varies with the thickness of the insulation, the shape and geometry of the electrodes, and the *rate* of application of voltage. Where no other information is specified, the puncture strength usually refers to that determined by tests using plane circular electrodes and applying voltage at a rate such that breakdown may occur in less than 2 or 3 minutes.

Two mechanisms of breakdown are of importance. Both may occur at the same time, although one or the other may be dominant. As mentioned in the previous paragraphs, under voltage stress, an increase in temperature causes an increase of the conductance (or a decrease in the resistance) of the insulating material. Conduction current flow implies a release of energy in the insulation. Under electrical stress near the puncture value, a local region may increase in temperature because heat is released faster than it is carried away. The increased temperature causes an increase in the conductance of that local area, and more current flows, thereby increasing the temperature even more. Of course, the increase of temperature will cause increased conduction of heat away from the region and a stable condition may result. However, if the voltage is further increased, the temperature may continue to rise and the current may continue to increase until chemical changes destroy the insulation and puncture results. This is sometimes called *thermal breakdown.*

Another mechanism of breakdown involves the acceleration of free electrons in an insulation volume to the energy level at which they, in collision with neutral molecules, knock other electrons free of the parent molecule, thus forming electron *avalanches*, as was discussed in Section 9-4c. In solids, these high-speed electrons break chemical bonds and so bring about chemical changes. In gases, the high-speed electrons generate ion pairs. The movement of the resulting charged particles, if sufficiently intense, produces sparkover and arcs. This may be termed breakdown by electron collision.

e. Volt–Time Characteristics of Breakdown. Both mechanisms of insulation breakdown described above require energy to be delivered from the voltage source to the insulation volume. This means that time is required for breakdown. The lower the applied voltage, the more slowly does the breakdown process proceed, and the longer the time to reach breakdown. At voltages below a critical value, the rate of energy dissipation by normal thermal conduction and convection processes is sufficient to prevent the growth of the breakdown process. At stresses below this critical value, insulation may operate indefinitely without breakdown.

Insulation is sometimes tested by applying a voltage above the critical value and observing the time which elapses before breakdown. Then a second experiment is performed with a different value of test voltage. This is repeated and finally a curve is drawn showing the time to breakdown plotted against applied voltage.

For solid insulation where the breakdown process is basically thermal, the curve shows a characteristic like that in Fig. 12-5a. Note that for voltages just above the critical value, V_c, the time lag of breakdown may be measured in minutes or even hours.

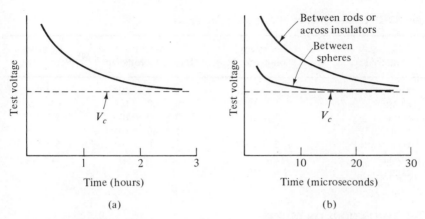

FIG. 12-5. Volt–time characteristics of insulation break-down.
(a) Solid insulation. (b) Air between spheres and between rods.

For gaseous insulation, where the breakdown process is basically by electron collision, the volt–time curve appears as shown in Fig. 12-5b. Note that the time here is measured in microseconds. The shape of the electrodes of the test system influences the shape of the curve. Between closely spaced spheres, where the electric field is almost uniform, the time lag is very small. Between rod gaps, where the electric field is very nonuniform, the time lag is greater.

    f. Insulation Deterioration. Insulation in service may deteriorate, a fact which implies that periodic checks of insulation condition must be made to avoid equipment failure due to breakdown of insulation.

    The conductivity of solid and liquid insulation increases rapidly with moisture content. Designs and maintenance procedures must be such as to minimize the amount of moisture that may be absorbed.

    Chemical changes due to excessive temperature and due to exposure to chemicals, sunlight, and nuclear radiation may permanently impair the value of solid and liquid insulation.

    Gaseous insulation, such as air, loses strength as particle density is decreased, either by a reduction of pressure or by an increase in temperature. Some gases, such as sulfur hexafluoride, lose insulation strength if contaminated with air or moisture.

    g. Typical Insulation Characteristics. The relative permittivity, puncture strength, and resistivity of several materials of importance in electric power are shown in Table 12-1. The puncture strength and resistivity will vary considerably, dependent on purity, manufacturing processes used, and moisture content. For comparison, the resistivity of copper is also shown.

Table 12-1. *Characteristics of Insulating Materials*

Material	Resistivity, ohm-cm	Puncture Strength, volts/cm	Relative Permittivity
AIR	Very high	3×10^4	1.0
GLASS	2×10^{13}	0.5–3.0×10^6	5.4–9.9
MICA	2×10^{17}	3.5–7.0×10^6	2.5–6.6
PAPER		1.0–4.0×10^5	2.0–2.6
PORCELAIN	3×10^{14}	1.0×10^5	5.7–6.8
RUBBER	1×10^{18}		2.0–3.5
COPPER	1.6×10^{-6}		

12-4. LIMITATIONS ON DESIGN

The insulation of an electric power system is of critical importance from the standpoint of service continuity. As mentioned before, probably more major equipment troubles are traceable to insulation failure than to any other cause. It might be argued that equipment should be overinsulated in terms of present practice. There are, however, other factors in addition to direct cost that argue against the use of higher insulation levels.

In cables, insulation is operated at very high stress. If insulation thickness were increased, more material would be required. In addition, larger-diameter cables would require more lead for covering, would be more difficult to handle, and lengths that could be put on reels would be reduced. In addition, electrical insulation is also good thermal insulation. Increased insulation thickness increases the problem of heat removal from the power conductors and requires a reduction of their current rating.

Increased thickness of insulation in transformers increases the size of coils and cores and increases copper and iron losses. The larger spacing between coils results in increased per unit impedance.

Increasing the number of suspension units in transmission line insulators necessitates an increase in cross-arm length, which in turn requires heavier structures and perhaps wider rights of way.

Similar statements could be made regarding other equipment, such as generators, instrument transformers, and circuit breakers. An arbitrary increase in insulation strength results in increased costs of associated parts and, in many instances, less satisfactory operating characteristics.

Because of the problems associated with equipment designs that attempt to utilize overly generous insulation, efforts are made in other directions. Manufacturers attempt to produce insulation of uniformly high quality, operators attempt to maintain the insulation with minimum deterioration, and designers attempt to plan systems in which overvoltages due to transient conditions are limited to values only slightly above the system operating voltage.

12-5. SOURCES OF OVERVOLTAGES

Overvoltages on power systems are traceable to three basic causes: lightning, switching, and contact with circuits of higher voltage rating. The power-system designer seeks to minimize the number of these occurrences, to limit the magnitude of the voltages produced, and to control their effects on operating equipment.

a. Lightning. Lightning results from the presence of clouds which have become charged by the action of falling rain and vertical air currents, a condition commonly found in cumulus clouds. Voltages may be set up on overhead lines due to *direct* strokes (Fig. 12-6a) and due to indirect

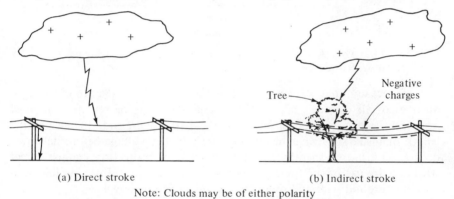

(a) Direct stroke   (b) Indirect stroke

Note: Clouds may be of either polarity

FIG. 12-6. Lightning stroke to a line. (a) Direct. (b) Indirect.

strokes (Fig. 12-6b). In a direct stroke, the lightning current path is directly from the cloud to the subject equipment—an overhead line in Fig. 12-6a. From the line, the current path may be over the insulators and down the pole to ground. The voltage set up on the line may be that necessary to flash over this path to ground.

In the indirect stroke, the current path is to some nearby object, such as the tree shown in Fig. 12-6b. The voltage appearing on the line is explained as follows: As the cloud comes over the line, the positive charges it carries draw negative charges from distant points and hold them bound on the line under the cloud in position as shown. The voltage on the line is zero. If the cloud is assumed to discharge on the occurrence of the stroke in zero time, the positive charges suddenly disappear, leaving the negative charges unbound. Their presence on the line implies a negative voltage with respect to ground.

On the occurrence of a stroke, lightning clouds do not discharge in zero time. Instead, the stroke current rises from zero value to maximum value (perhaps 50,000 amperes) in a few microseconds and is completed in a few hundred microseconds. For many test and design purposes, the voltage set up on the line due to lightning is assumed to be as shown in Fig. 12-7. Here it may be noted that the crest value of the voltage is reached in 1.2 microseconds and the 50 percent point on the tail of the wave is reached in 50 microseconds. Obviously all voltages set up by natural lightning discharges do not conform to this specification.

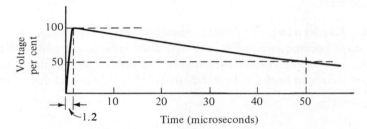

FIG. 12-7. A 1.2 × 50 test wave used to simulate a lightning-produced voltage.

Direct lightning strokes to lines as shown in Fig. 12-6a are of concern on lines of all voltage class, as the voltage that may be set up is in most instances limited by the flashover of the path to ground. Increasing the length of insulator strings merely permits a higher voltage before flashover occurs. The most generally accepted method of protection against direct strokes is by use of the *overhead ground wire* (Fig. 12-8). For simplification only one ground wire and one power conductor are shown.

The ground wire is placed *above* the power conductor at such a position that practically all lightning-stroke paths will be to it instead of to the power conductor. Stroke current then flows to ground, most of it passing through

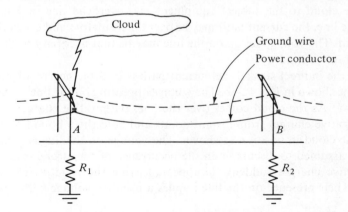

FIG. 12-8. A lightning stroke to a transmission line with overhead ground wire.

the tower footing ground resistance R_1 while a smaller part goes down the line and to ground through the adjacent tower footings. The tower rises in voltage due to the current I_1 through the resistance R_1 to a value which is (as a first approximation)

$$V_t = I_1 R_1 \tag{12-2}$$

Approximately this voltage appears between the tower and the power conductor (which was not struck). If this voltage is less than that required to cause insulator flashover, no trouble results. Protection by this method is improved by using two carefully placed ground wires and by making tower footing ground resistance of low value. Ground resistance is discussed in Chapter 13.

> EXAMPLE 12-1. Suppose the lightning current which flows to ground through the tower footing is 25,000 A. If the tower footing resistance is 10 Ω, using Eq. (12-2), the potential of the tower rises to 250,000 V. This is the approximate voltage across the insulator string. However, if the ground resistance is 25 Ω, the voltage is 625,000 V, and insulator flashover will be more probable.

The lightning record of lines supported on towers 80 to 90 feet tall substantiates the simple theory of line protection just presented. The poorer record of lines on towers over 100 ft in height indicates that other factors, perhaps the inductance of the path down the tower, should be considered.

Indirect strokes produce relatively low voltages on lines. They are of real concern on low-voltage lines supported on small insulators. They are of little importance on high-voltage lines whose insulators can withstand hundreds of kilovolts without flashover.

b. Switching Transients. Overvoltages may result from switching operations on the closing of circuit breakers, or if restriking occurs within the breaker, on circuit-opening operations. A simple circuit to explain the nature of switching transients is shown in Fig. 12-9. In this circuit L

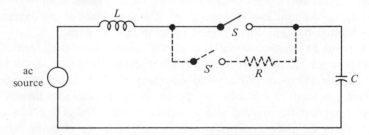

FIG. 12-9. A circuit in which overvoltages may be produced by switching.

might represent the inductance of a transformer while C represents the capacitance of a transmission line. When the switch S is closed, oscillations will result in a voltage of frequency

$$f = \frac{1}{2\pi\sqrt{LC}} \qquad\qquad (12\text{-}3)$$

appearing across the capacitor. The magnitude of this voltage depends on the magnitude of the voltage left on the capacitor due to previous operation, the voltage of the ac source, and the point on the wave at which the switch makes contact. The magnitude of the overvoltage may be controlled by (1) causing the switch to close at the most favorable point on the wave, or (2) inserting resistance in the circuit by closing an auxiliary switch S' (shown dashed) a short interval of time before the main contact S closes. Both methods are in use on major power systems.

c. **Contact With Circuits of Higher Voltage.** Lines of different voltage class are sometimes carried on the same structures. Necessary crossings of lines of different voltage class bring these lines in close proximity to each other. Under these circumstances, conductor or support failure or extreme loading due to wind and sleet may bring lines in contact with each other. In practice the higher voltage line is placed above the lower voltage line, and because of its greater importance, the higher voltage line may be more liberally designed. However, contact between circuits does occasionally occur. Such contacts are recognized by the relays as faults, and circuit breakers are opened to clear the trouble. For short periods of time the lower-voltage line may be subjected to overvoltage.

Contact between circuits of different voltage class may also result from failure of the winding-to-winding insulation in transformers. Here, again, the trouble should be recognized by the protective system and the faulted equipment removed by operation of circuit breakers or fuses.

During the short interval of time between the establishment of contact between circuits and the removal of the defective equipment, an overvoltage may be imposed on the low-voltage circuit. The magnitude of this overvoltage may be limited by lightning arresters or by protective gaps.

A point of extreme concern is at the distribution transformer that supplies residential load. Here, flashover of the transformer bushings or failure of the insulation between primary and secondary windings might cause primary voltage (perhaps 12 kilovolts) to appear on the circuits into homes, thus creating a serious fire hazard and a danger to the lives of the occupants. This problem is quite effectively solved by providing primary lightning arresters and by grounding them, the transformer case, and the secondary neutral at the transformer. A ground connection is also placed on the secondary neutral on the customer's premises. Principally by the ground connections, the danger of serious overvoltages is greatly reduced. An accidental connection between primary and secondary circuits results in a ground fault on the primary and the "blowing" of the primary fuses.

The grounding of the secondary neutral of residential distribution circuits greatly reduces the overall hazard to people and property. However, a source of danger still exists which is not generally recognized and which accounts for a few deaths each year. Figure 12-10 shows a typical distribution

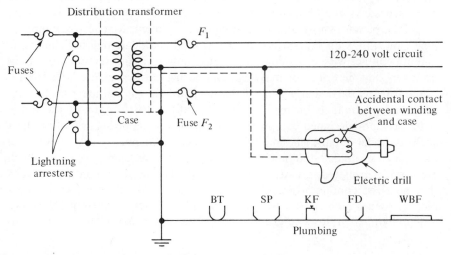

FIG. 12-10. A typical residential distribution circuit showing equipment which is directly connected to the circuit neutral.
BT   Bathtub
SP   Swimming Pool
KF   Kitchen Faucet
FD   Furnace Duct
WBF   Wet Basement Floor.

transformer supplying a 120–240 volt circuit into a home. The secondary neutral, being tied to ground, is also directly connected through the grounded piping of plumbing fixtures and furnaces to such equipment as bathtubs (BT), swimming pools (SP), kitchen faucets (KF), furnace ducts (FD), and wet basement floors (WBF). A person in contact with any of these is tied directly to one side of a 120-volt circuit. If at the same time he touches the other (or hot) side of the circuit, he is directly across the supply. A contact with the hot side of the circuit may be made with a metal floor lamp or metal receptacle in which an unnoticed wiring defect exists. Figure 12-10 illustrates a situation that occurred near the home of the writer. A man used an electric drill in which there was an accidental contact between one terminal of the drill motor and the metal case. Unfortunately he chose to use the (defective) drill while standing in a swimming pool. When he pulled the trigger of the drill to close the motor switch, the case of the drill was then connected directly to the hot side of the 120-volt circuit. The man was electrocuted.

The importance of ground leads on portable tools and appliances may be illustrated from the above example. Had there been such a ground lead (shown dashed) on the (defective) drill the closing of the switch would have produced a short circuit on the 120-volt circuit. This trouble would have been cleared by the "blowing" of fuse F_2 without injury to the workman.

The person who works on grounded equipment, touches plumbing or pipes, or stands on the surface of the ground should remember that he is making contact with one side of an electric power circuit. Great care should be taken that he does not make contact with the other side of the circuit. The hazard is particularly great if the hands of the person are wet, for wet skin readily conducts current into the liquid-flooded interior of the body.

Specially designed equipment is available for the protection of workers such as boiler repairmen, miners, and others who must work in intimate relation to grounded equipment. Portable tools have been developed which are doubly insulated, greatly reducing the probability of a contact between the energized conductor and the hands of the workman. Another arrangement provides high-speed opening of the power circuit if current (of even a few milliamperes) flows to ground rather than returning in the neutral conductor.

12-6. TRAVELING WAVES

As described in Section 12-5a, the voltages set up on lines by lightning rise from zero to maximum value in a few microseconds. Transients due to switching produce voltages that rise very rapidly, although not as fast as those due to lightning. These voltages may be transmitted many miles over lines and may produce overvoltages in stations far from their origin. The propagation of an electrical disturbance down a line is very rapid but at a measurable rate. When very rapid voltage changes are involved, it is helpful to look at the circuit behavior in terms of *traveling waves* rather than by conventional methods of analysis.

A transmission line (overhead or cable) has a certain inductance per unit length L and certain capacitance per unit length C. For this analysis, resistance and leakage will be neglected. The inductance and the capacitance are distributed all along the line. Each section, even a very short length of line, possesses both L and C. If the line is divided up into small sections of length Δx, the inductance of that section will be $L\Delta x$ and the capacitance will be $C\Delta x$. A line which is divided into small sections in each of which the inductance and capacitance is lumped is shown in Fig. 12-11. The length Δx may be made as short as desired; thus uniform distribution of the line constants may be represented as closely as desired.

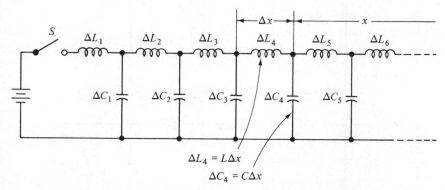

FIG. 12-11. A transmission line shown as incremental sections of inductance and capacitance.

Suppose the switch S is closed at time $t = 0$. The nature of inductance is such that time is required to establish a current of significant magnitude through the inductor ΔL_1. After this current is established, time will be required to build up a voltage on the capacitor ΔC_1. Only after voltage appears here will current start building in inductor ΔL_2. It is obvious that the conditions which exist at the source end of the line are not immediately observed at the far end of the line. The propagation of the source-end voltage and current conditions down the line is termed a *traveling wave*. The velocity of propagation, the relation between current and voltage at any point, the energy transmitted, the energy stored in a section of line, and other relations may be determined from two simple differential equations, derived below.

In Fig. 12-12a we may examine the voltages which exist in an incremental section. At distance x from the far end of the line, assume that the voltage is v and the current is i. At distance $x + \Delta x$ the voltage is $v + \Delta v$. The value of Δv may be written

$$\Delta v = L\,\Delta x \frac{di}{dt} \tag{12-4}$$

In Fig. 12-12b we may examine the currents which exist in an incremental section:

$$\Delta i = C\,\Delta x \frac{dv}{dt} \tag{12-5}$$

If Δx is made very short, these equations take on the differential form:

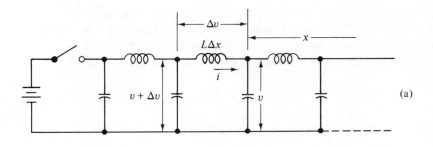

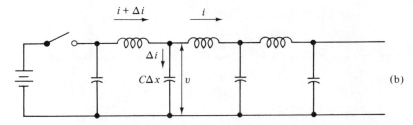

FIG. 12-12. An incremental section of a line. (a) Voltage conditions. (b) Current conditions.

$$\frac{\partial v}{\partial x} = L\frac{\partial i}{\partial t} \qquad (12\text{-}6)$$

$$\frac{\partial i}{\partial x} = C\frac{\partial v}{\partial t} \qquad (12\text{-}7)$$

These are the basic equations, from which much can be learned about traveling waves.

The relation between voltage and current in a traveling wave may be determined by the following steps. Dividing Eq. (12-6) by Eq. (12-7) yields

$$\frac{\partial v}{\partial i} = \frac{L}{C}\frac{\partial i}{\partial v}$$

Then

$$\frac{(dv)^2}{(di)^2} = \frac{L}{C}$$

$$\frac{dv}{di} = \sqrt{\frac{L}{C}}$$

or

$$dv = \sqrt{\frac{L}{C}}\, di$$

and, on integrating,

$$v = \sqrt{\frac{L}{C}}\, i \tag{12-8}$$

$$v = Z_c i \tag{12-9}$$

where

$$Z_c = \sqrt{\frac{L}{C}} \tag{12-10}$$

is termed the characteristic impedance of the line. Note that it is a constant relating v and i, and so has the character of a resistor.

The *velocity* of propagation U may be determined from the relation

$$U = \frac{dx}{dt}$$

From Eq. (12-6),

$$\frac{\partial x}{\partial t} = \frac{1}{L}\frac{\partial v}{\partial i}$$

From Eq. (12-7),

$$\frac{\partial v}{\partial i} = \frac{1}{C}\frac{\partial t}{\partial x}$$

Then

$$\frac{dx}{dt} = \frac{1}{LC}\frac{dt}{dx}$$

$$\frac{(dx)^2}{(dt)^2} = \frac{1}{LC}$$

$$U = \frac{dx}{dt} = \frac{1}{\sqrt{LC}} \tag{12-11}$$

The power P flowing past any point is

$$P = vi = \frac{v^2}{Z_c} = i^2 Z_c \tag{12-12}$$

The energy stored in the inductance of the line, per unit length, is

$$W_L = \tfrac{1}{2} L i^2$$

The energy stored in the capacitance of the line, per unit length, is

$$W_C = \tfrac{1}{2}Cv^2$$

The total energy stored in the line, per unit length, is

$$W = W_L + W_C = \tfrac{1}{2}Li^2 + \tfrac{1}{2}Cv^2$$

Since

$$v = i\sqrt{\frac{L}{C}}$$

$$v^2 = i^2\frac{L}{C}$$

and

$$W = \tfrac{1}{2}Li^2 + \tfrac{1}{2}Li^2$$
$$= Li^2 \qquad\qquad (12\text{-}13)$$

Similarly,

$$W = Cv^2 \qquad\qquad (12\text{-}14)$$

If typical values are substituted into Eqs. (12-10) and (12-11), it is found that the characteristic impedance and velocity of propagation on overhead lines and cable circuits are approximately as given in Table 12-2.

Table 12-2. *Characteristics of Overhead and Cable Lines*

	Characteristic Impedance Z, ohms	Velocity of Propagation U miles/sec	feet/μ sec
OVERHEAD LINE	400	186,000	1000
CABLE	50	90,000	500

Let us examine the time behavior of the overhead line shown in Fig. 12-11 when it is energized by closing the switch S, connecting it to a constant 2000-volt source. This line is shown redrawn in Fig. 12-13a, simplified by the omission of the inductors and capacitors. This line has a characteristic impedance of 400 ohms, a velocity of propagation of 1000 feet per microsecond, and is assumed to be 5000 feet in length.

Suppose now that the switch S is closed at $t = 0$. A voltage and current profile along the line will appear as in Fig. 12-13b. A voltage and current

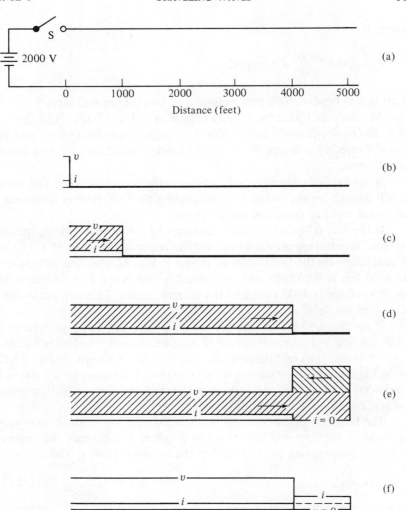

FIG. 12-13. A transmission line being energized by switching on a
constant voltage source. (a) The circuit arrangement. (b) Voltage and
current conditions at $t = 0+$. (c) Voltage and current conditions
at $t = 1$ μsec. (d) Voltage and current conditions at $t = 4$ μsec.
(e) Voltage and current conditions, line open circuited at terminal,
$t = 6$ μsec. (f) Voltage and current conditions, line short circuited
at terminal, $t = 6$ μsec.

exists at the right-hand terminal of the switch, but at other points on the
line the voltage and current are zero.

At time $t = 1.0$ μ sec, the voltage and current condition on the line has
propagated 1000 feet and the profiles appear as in Fig. 12-13c. In the line
section from 0 to 1000 feet, the voltage is at all points 2000 volts and the

current is

$$i = \frac{v}{Z_c} = \frac{2000}{400} = 5 \text{ amperes}$$

At all points beyond 1000 feet, voltage and current are still zero.

At time $t = 4\ \mu$sec the profiles appear as in Fig. 12-13d. Now the voltage is 2000 volts from S to the 4000-foot point, and the current over this same distance is 5 amperes. Beyond the 4000-foot point, voltage and current are still zero.

What happens when the wave comes to the end of the line? The behavior will depend on the method of terminating the line. Energy continues to flow down the line from the power source.

If the line is open-circuited, there can be no current flow at the end, and so all received energy must be stored in the line capacitance. At $t = 6\mu$ sec, the conditions on the line will be as shown in Fig. 12-13e. The voltage from 0 to 4000 feet is 2000 volts and the current is 5 amperes. From 4000 to 5000 feet, the voltage is 4000 volts and the current is zero. This new condition is moving to the left.

If the line is short-circuited at the end, there can be no voltage at the end of the line and all received energy must be stored in the line inductance. At $t = 6\ \mu$ sec, the conditions on the line will be as shown in Fig. 12-13f. The voltage from 0 to 4000 feet is 2000 volts and the current is 5 amperes. From 4000 to 5000 feet, the voltage is zero and the current is 10 amperes. The new condition is moving to the left.

The behavior of the line may be analyzed as if an incident wave propagates down the line and reflects when it comes to the end, the reflected wave then propagating to the left. For the incident wave, v_i and i_i,

$$v_i = Z_c i_i \tag{12-9}$$

and for the reflected wave, v_r and i_r,

$$v_r = -Z_c i_r \tag{12-15}$$

(Note that in the reflected wave, the current is flowing to the left, hence the negative sign.)

The voltage v and the current i where both waves exist are

$$v = v_i + v_r \tag{12-16}$$

$$i = i_i + i_r \tag{12-17}$$

where v_i and i_r apply to the incident waves and v_r and i_r apply to the reflected waves.

Refer to Fig. 12-14a, which shows a line terminated by a resistor R.

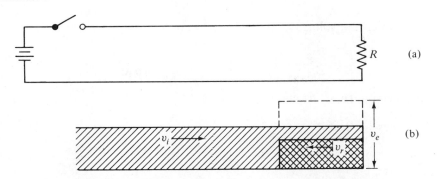

FIG. 12-14. A line being energized by a constant voltage source. The line is terminated by a resistor R. (a) The circuit arrangement. (b) Voltage conditions on the line after reflection.

In Fig. 12-14b an incident voltage wave v_i and a reflected voltage wave v_r are shown. The power carried to the right by the incident wave is

$$P_i = \frac{v_i^2}{Z_C} \tag{12-18}$$

The power carried to the left by the reflected wave is

$$P_r = \frac{v_r^2}{Z_C} \tag{12-19}$$

The voltage v_e on the end of the line at the resistor is

$$v_e = v_i + v_r$$

and the power P_R absorbed by the resistor is

$$P_R = \frac{v_e^2}{R} = \frac{(v_i + v_r)^2}{R} \tag{12-20}$$

Then

$$P_i = P_r + P_R \tag{12-21}$$

$$\frac{v_i^2}{Z_C} = \frac{v_r^2}{Z_C} + \frac{(v_i + v_r)^2}{R}$$

$$\frac{v_i^2 - v_r^2}{Z_C} = \frac{(v_i + v_r)^2}{R}$$

$$\frac{(v_i - v_r)(v_i + v_r)}{Z_C} = \frac{(v_i + v_r)^2}{R}$$

$$\frac{v_i - v_r}{Z_C} = \frac{v_i + v_r}{R}$$

from which the voltage of the reflected wave is

$$v_r = v_i \frac{R - Z_C}{R + Z_C} \tag{12-22}$$

and the voltage at the end of the line is

$$v_e = v_i + v_r$$
$$= 2v_i \frac{R}{R + Z_C} = 2v_i \frac{1}{1 + Z_C/R} \tag{12-23}$$

This equation has the form of the expression for the voltage across R in the circuit of Fig. 12-15.

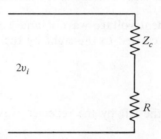

FIG. 12-15. The circuit equivalent of a line terminated in a resistor.

The current through the resistor at the end of the line is

$$i_e = \frac{v_e}{R} = \frac{2v_i}{R + Z_C} \tag{12-24}$$

Suppose the line of Fig. 12-13 is terminated with a resistance of 600 ohms. When the 2000-volt, 5-ampere incident wave reaches the end, a reflection will occur. The voltage of the reflected wave will be, by Eq. (12-22),

$$v_r = v_i \frac{R - Z}{R + Z} = 2000 \frac{600 - 400}{600 + 400}$$
$$= 400 \text{ V}$$

The voltage at the end of the line will be, by Eq. (12-23),

$$v_e = 2v_i \frac{R}{R + Z} = 2 \times 2000 \frac{600}{600 + 400} = 2400 \text{ V}$$

The current in the reflected wave is, by Eq. (12-15),

$$i_r = \frac{-v_r}{Z_C} = -\frac{400}{400} = -1.0 \text{ A}$$

The current through the resistor is by Eq (12-24)

$$i_R = \frac{v_e}{R} = \frac{2400}{600} = 4 \text{ A}$$

Note that if the resistance R at the end of the line is less than Z_C, the reflected voltage wave will be negative and the reflected current wave will be positive.

This mathematical analysis is consistent with the behavior described in Fig. 12-13e and f. If R is infinite (open circuit), by Eq. (12-23), the voltage at the end of the line becomes

$$v_e = 2v_i \frac{1}{1 + Z_C/R} = 2v_i$$

If R is zero,

$$v_e = 2v_i \frac{R}{R + Z_C} = 0$$

Several facts of immediate importance appear from the knowledge of traveling waves here presented.

1. Refer to Fig. 12-8. The effect of current flow to the ground at resistance R_2 is not felt until sufficient time has elapsed for a wave to travel from tower A to B and return. With long spans, this may be several microseconds. This may allow sufficient time for the lightning-produced voltage to rise to a value sufficiently high to cause flashover of the insulators at tower A.

2. Reference to Eq. (12-23) shows that if a line is open at the end ($R = \infty$), the voltage at that point will be $2v_i$. Hence terminal

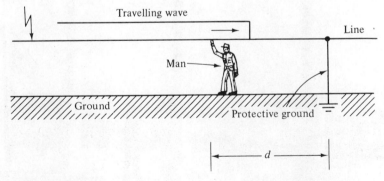

FIG. 12-16. A man working on a line between his protective grounds and the source of a traveling wave.

equipment on an open-ended line may experience a voltage higher than that of the initial traveling wave which produced it.

3. A set of grounds put on a line to protect workmen from lightning-produced voltages must be placed either at the location where the men are working or between the workmen and the lightning source. If placed as shown in Fig. 12-16, the workmen will be subjected to full incident wave voltage for the period required for the wave to travel from the men to the ground and return, a distance $2d$.

12-7. LIGHTNING ARRESTERS

Electric power equipment is subjected to overvoltages from various causes, and the equipment must have adequate insulation to avoid failure. If the overvoltages can be limited to values only slightly greater than the normal operating voltage, favorable designs may be attained. The effects of excessive insulation were discussed in Section 12-4.

Lightning arresters Fig. 12-17 are devices put on electric power equipment to limit overvoltages to a value less than they would be if the arresters were not present. Ideally a lightning arrester should be off the line under normal operation, switch onto the line when the voltage is perhaps 20 per-

FIG. 12-17. A set of 138 kV lightning arresters.

cent above normal value, limit the voltage to this value regardless of the nature or source of the overvoltage, and switch off of the line when the disturbance is past and normal voltage has been restored.

The basic form of a lightning arrester is shown in Fig. 12-18. A spark

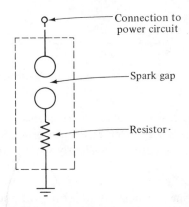

Connection to power circuit

Spark gap

Resistor

FIG. 12-18. A basic lightning arrester.

gap is connected in series with a resistor. The gap is set at a sparkover value greater than normal line voltage, hence the gap is normally non-conducting. On the occurrence of an overvoltage, the gap sparks over, and then the voltage across the arrester terminals is determined by the IR drop in the arrester. The resistor limits the current flow, avoiding the effect of a short circuit. When the overvoltage condition has passed, the arc in the gap should cease, thus disconnecting the arrester from the circuit. If the arc does not go out, current continues to flow through the resistor, and both the resistor and the gap may be destroyed.

The ohmic value of the resistor is of importance. If it is low, the voltage across it is low during the flow of transient current through it and the device functions effectively as a protection against overvoltage. However, the power follow current through the gap will be high and will be difficult to interrupt. The arrester shown in Fig. 12-18 has the disadvantage that the voltage across it increases linearly with the current flow through it. Thus severe lightning transients will cause proportionately high voltages across the arrester terminals. Nonlinear resistors have been developed for lightning-arrester use. These resistors have current–voltage characteristics somewhat as shown in Fig. 12-19. With this behavior, it may be noted that the resistance drops as voltage (or current) is increased. Suppose a line is terminated in an arrester. Equation (12-23) gives the voltage across its terminals as

$$v_a = 2v_i \frac{R}{R + Z_c} \qquad (12\text{-}23)$$

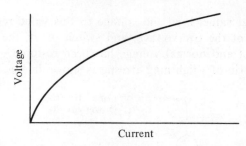

FIG. 12-19. The volt–ampere relationship in the non-linear resistor of an arrester.

It may be noted that as R is decreased, v_a becomes a smaller fraction of v_i. Hence an arrester with a nonlinear resistance element permits across its terminals a voltage that increases only slightly with an increase of the value of the incident wave.

As pointed out in a previous section, the arrester resistor must be disconnected from the line immediately following the end of the overvoltage condition. The interruption of the power follow current by the gap presents the same basic problems as the interruption of the arc in an opening circuit breaker. In one respect the problem is more difficult. In the arrester the arcing contacts are fixed in position, rather than moving apart as in the circuit breaker. In another respect the problem is less difficult. Although the nonlinear resistor of the arrester passes high-magnitude currents during periods of overvoltage, the current with normal voltage across the arrester terminals is proportionately smaller. Hence the power-frequency currents to be interrupted are much smaller than in a circuit breaker.

Figure 12-18 showed the basic elements of a lightning arrester. For application on high-voltage circuits, a large number of these basic elements may be put in series to form the arrester. This design has the advantage that standard parts may be used for arresters of different voltage rating. It also has the advantage that arc interruption is done through a series of gaps, an arrangement discussed in Chapter 9. The spark gaps used in the arresters are designed to have small time lag, uniform breakdown voltage, and favorable arc-extinction characteristics.

A lightning-arrester spark gap of one design is shown in Fig. 12-20. Two diverging metal electrodes A and B are mounted on a disc of vitreous material such as porcelain. Another piece of similar material fits closely on top of the electrodes. Sparkover occurs across the small gap G. The arc moves out progressively into positions 1, 2, and 3. This movement is assured by a magnetic field set up vertically in the arc space by the flow of arrester current through coils placed above and below the gap. The narrow slot between the porcelain supports decreases in thickness toward arc position 3. Arc-extinguishing forces are then the lengthening of the arc and the deioniz-

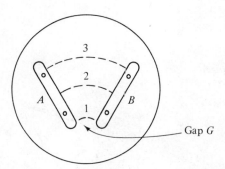

FIG. 12-20. The confined spark gap of a lightning arrester.

ing effect of the arc confined to the narrow region between the porcelain discs. In some instances arc extinction occurs before the first zero of the power-frequency current.

Arresters must be placed very near the equipment to be protected. In many instances arresters are mounted directly on the tanks of large power transformers. If placed at a distance from the equipment to be protected, traveling-wave conditions may result in a voltage at the equipment much higher than that permitted at the arrester.

12-8. TRANSMISSION-LINE INSULATION

The conductors of overhead transmission lines Fig. 12-21 are supported by porcelain insulators and are insulated from each other by air between the points of attachment.

Modern porcelain insulators are designed and manufactured in such a fashion that in themselves they are almost perfect in operation. Very seldom is porous or cracked porcelain found. Flashover of line insulators is almost always traceable to the breakdown of the air around them due to overvoltage from lightning or other causes. Insulators whose surfaces are contaminated and then moistened by light rain or fog may flash over even under normal-operating-voltage conditions. If an insulator is cracked or porous and permits lightning or power-frequency current to pass through the body of the insulator, it may be shattered, with the resultant dropping of the line.

The air between the conductors of a high-voltage transmission line is under electrical stress. This stress is relatively great immediately adjacent to the conductors and very low midway between them. When the stress in the air exceeds about 30 kilovolts/cm, breakdown occurs within that area where the high stress exists. Hence on a transmission line it is possible to have dielectric breakdown of the air around the conductors without total breakdown between conductors. This condition is termed *corona.*

FIG. 12-21. A 765 kV transmission line showing the ground wires, bundle conductors, and *V*-string insulators.

Corona on transmission lines produces power loss, generates ozone and acid compounds of nitrogen, and produces radio interference and audible noise. These effects are easily tolerated if of low level but can become very annoying if excessive. A great amount of experimental work has been done to study these effects, for they present limiting factors in the voltage at which lines may be operated. Present-day designs permit these effects but attempt to control their levels to a point where they are relatively unobjectionable.

Problems

12-1. Two parallel plates each 20 by 20 cm are separated by a distance of 0.35 cm. What will be the capacitance between the plates if the dielectric between them is

a. air?
b. glass?

12-2. Refer to Problem 12-1. What will be the approximate value of the voltage necessary to cause puncture or breakdown of the insulation?

12-3. Refer to Problem 12-1. Assume that the insulation is glass. With 25,000 volts dc between the plates, what will be the leakage current and the power loss in the insulation? Will the power loss be greater or less if the test voltage is alternating? Explain.

12-4. A 12-kilovolt power-factor correction capacitor bank has been removed from service by opening a circuit breaker. Consider the point on the wave where the breaker current is interrupted and determine the value (if any) of the voltage that might remain on the capacitor.

12-5. An underground cable connected to an overhead line is taken out of service for maintenance. A safety ground connection is put on the line. Next the cable is disconnected from the line. Describe any hazard which may exist.

12-6. Compare the mechanism of breakdown in the puncture of transformer insulation and in the breakdown of an air gap.

12-7. A transformer bushing has a protective air spark gap in parallel with it. On test the voltage is raised to a value well above that at which this equipment is designed to operate and although held for 30 seconds, no breakdown has occurred. If the test is continued at constant voltage and breakdown later occurs, which piece of equipment, the bushing or the spark gap, would you expect to be the one that failed? Explain.

12-8. For test, a standard wave (Fig. 12-7) is impressed on a power circuit. As the wave travels down the line, how many feet will it be (at any instant) from the point on the line at which the voltage first starts to rise and the point at which the voltage is a maximum? Assume that circuit to be

a. an overhead line.
b. a cable.

12-9. On the basis of the calculated inductance and capacitance per mile line-to-neutral, determine the characteristic impedance and velocity of propagation on a line consisting of 500,000 cir mil copper spaced 12 feet horizontally.

12-10. What is the characteristic impedance and velocity of propagation of a coaxial cable with rubber insulation? The diameter of the center conductor is 0.8 cm and that of the outer conductor is 2.1 cm.

12-11. The insulators on a transmission line will flash over at 600 kilovolts. In a section of this line the tower footing ground resistance is 18 ohms. What minimum value of lightning stroke current to the ground wire will result in insulator flashover?

12-12. Refer to Problem 12-11. If the stroke contacts the power conductor itself, what minimum value of current will cause flashover? (*Hint:* Consider the stroke current to the line to divide, setting up two traveling waves that flow in opposite directions from the point of contact.)

12-13. An overhead line is terminated by a resistor of 250 ohms. A lightning-produced traveling wave has a voltage of 400 kilovolts. Determine the power in the incident wave, in the reflected wave, and the power absorbed by the resistor. Consider $Z_C = 400$ ohms.

12-14. A long overhead line ($Z_C = 400$ ohms) connects to a long cable ($Z_C = 50$ ohms). A 250-kilovolt traveling wave originates on the line and propagates toward the junction. What will be the voltage at the junction? (*Note:* A long transmission circuit looks like a simple resistor as far as traveling waves are concerned.)

12-15. Repeat Problem 12-14 but assume that the 250-kilovolt traveling wave originates in the cable.

12-16. Describe four locations in your own experience where use of an ungrounded power tool would be particularly hazardous.

chapter 13

Grounding
Electric Circuits
Effectively

13-1. GROUNDING PRACTICE

In many types of electrical systems it is necessary to install in the ground electrodes or systems of electrodes, which serve as electrical connections to the earth. Such a connection is ordinarily referred to as a *ground*. Any device that is directly connected to a ground is said to be *grounded*. Grounds are required on power systems, on radio broadcast stations, on telephone systems, on lightning rods, and on the steel frames of large buildings.

Many of the devices associated with an electric power system are grounded. In residences, the entrance box and the secondary neutral are grounded. Ground terminals are provided in many of the convenience outlets which may be used to supply portable tools. In a residence, the ground consists of a connection to the water system, if it is available. In some instances, wires embedded in the footings of the building provide a connection to the earth. Driven rods are sometimes required.

At the transformers supplying distribution secondary circuits, a ground connection is provided. To this is connected the secondary neutral and in many instances the transformer case and the lightning arresters. Grounds are provided on the handles that operate airbreak disconnect switches and on regulators and similar pieces of equipment. Transmission lines with overhead ground wires have a ground connection at each supporting structure to which the ground wire is connected.

In generating and switching stations the frames of all generators and motors, the tanks of transformers and oil circuit breakers, and the cases of instruments and relays are grounded. The steel frames of buildings, the steel supporting structures in switchyards, and the fences which surround the stations are all grounded. By having all these parts connected to the same ground, the possibility of a dangerous voltage difference between parts is virtually eliminated.

Grounds are also provided at many points which form part of the electric circuits themselves. The neutrals of wye-connected generators and transformers may be solidly connected to ground or connected to ground through an impedance. Lightning arresters and shunt reactors are solidly grounded. The secondaries of current transformers and potential transformers have a ground at one point on each of the circuits.

The ground-electrode system of a generating or switching station may consist of an extensive network of buried conductors and driven ground rods. Great care is taken in the design and construction of these ground mats to make certain that their resistance is but a small fraction of an ohm.

In normal operation, the electrical connections to the earth (or grounds) carry currents of very low magnitude. These continuous small currents are frequently traceable to minor unbalances which exist in the transmission and distribution lines of the system. The electrical connections to the earth are of primary importance during those short intervals of time when faults exist on the system or when overvoltages arise due to lightning or switching transients. During these *abnormal* periods, which are of short duration, the current conducted through the ground connections may be of magnitudes measured in many thousands of amperes.

Ideally, the resistance of a ground should be zero. Then the flow of current through it causes no rise in the potential of the ground point measured relative to other points on the earth distant from the ground connection. It is, of course, impossible to install a ground-electrode system in the earth which has zero resistance, even with unlimited expenditure of material and effort. It is difficult to define a particular ohmic value of the ground connection which may be considered adequate; a lower value is always preferred. If the resistance to ground is finite (and it always is), current flow through the ground connection results in (1) a rise in potential of the equipment tied to the ground point measured relative to contacts with the earth at distant points, and (2) electric potential gradients measured horizontally on the earth's surface particularly in close proximity to the ground-electrode system. Both results present a shock hazard to personnel and an insulation damage hazard to equipment. These hazards are reduced as the resistance of the ground connection is lowered, but they can never be entirely eliminated.

13-2. GROUND RESISTANCE

The physical structure of a ground-electrode system may take many different forms. It may be a buried plate or a driven rod; it may be a group of two or more driven rods or a system of buried horizontal wires; it may be a combination of buried wires and driven rods. The concept of ground resistance may be demonstrated by considering the simplest form of an electrode, a metal hemisphere buried flush with the surface of the earth (Fig. 13-1).

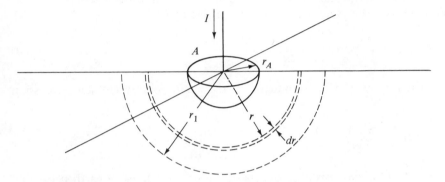

FIG. 13-1. A hemispheric metal electrode of radius r_A buried flush with the surface of the earth.

Assume that a current I from some source (not shown) enters this electrode and flows out radially into the surrounding ground, which is assumed to be of uniform resistivity and of infinite extent. The resistance of this ground connection may be calculated by reference to the well-known relation

$$R = \rho \frac{l}{A} \tag{13-1}$$

in which R is the resistance between two faces of a conducting body of resistivity ρ, length l, and of uniform cross-sectional area A at right angles to the current flow. In the case of the current path in the ground, its cross-sectional area is *not* uniform but increases with increasing distance from the electrode center. The cross-sectional area at right angles to the current flow may be seen to be that corresponding to the area of a hemisphere of radius r. Hence

$$A = 2\pi r^2 \tag{13-2}$$

Since this cross-sectional area applies only for the length dr (Fig. 13-1) of the current path, Eq. (13-1) must be modified to

$$dR = \rho \frac{dr}{2\pi r^2} \tag{13-3}$$

The incremental resistance dR may be interpreted as the resistance between opposite faces of a thin hemispheric shell of radius r and thickness dr. The resistance between the metal hemisphere A of radius r_A and an imagined hollow metal hemispheric electrode of radius r_1 is obtained by summing the resistances of an infinite number of thin shells which might be placed between A and the large hemisphere:

$$R = \int dR = \int_{r_A}^{r_1} \rho \frac{dr}{2\pi r^2} \tag{13-4}$$

When this expression is integrated and limits substituted, an expression results

$$R = \frac{\rho}{2\pi}\left(\frac{1}{r_A} - \frac{1}{r_1}\right) \tag{13-5}$$

This, then, is the resistance of a path through a volume of earth between a hemispheric electrode of radius r_A and a larger one of radius r_1.

EXAMPLE 13-1. Suppose a metallic hemisphere of radius 25 centimeters is buried flush with the surface of soil whose resistivity is 5000 ohm-centimeters. Calculate the resistance between this hemisphere and a hemispheric metal shell of radius 250 centimeters.

Solution. Substituting in Eq. (13-5) yields

$$R = \frac{5000}{2\pi}\left(\frac{1}{25} - \frac{1}{250}\right)$$

$$= \frac{5000}{2\pi}(0.040 - 0.004) = 28.6 \text{ ohms}$$

If the radius r_1 in Eq. (13-5) is made very great, the equation reduces to

$$R = \frac{\rho}{2\pi r_A} \tag{13-6}$$

Equation (13-6) gives a physical interpretation of the meaning of the ground resistance of an electrode buried in the earth. The ground resistance of a hemispheric electrode (or any other form of electrode that may be chosen) is the resistance from the metallic electrode itself through a path of ever-increasing cross section to an imagined hemispheric electrode of infinite radius. For practical considerations, it is not necessary to have an electrode of infinite radius as just described. Referring to Eq. (13-5), it may be noted that if r_1 is 10 times the value of r_A, the calculated resistance R is 90 percent of the value given by Eq. (13-6). From this it may be concluded that most of the resistance of an earth connection resides in that soil volume which is very close to the ground electrode.

13-3. EXPRESSIONS FOR GROUND RESISTANCE

An expression for the ground resistance of a buried hemispheric electrode was developed in Section 3-2. The mathematics for this development was relatively simple, but a hemispheric electrode shape is seldom used in practice. Expressions for the resistance of practical electrode shapes have been derived from field theory, a subject beyond the present discussion. The various equations for ground resistance of a number of commonly used electrode shapes are presented in Table 13-1.

> EXAMPLE 13-2. Using the equation given in Table 13-1, calculate the resistance of a single ground rod of length 10 feet (30.5 centimeters) and radius $\frac{1}{2}$ inch (1.27 centimeters) in soil of resistivity 200,000 ohm centimeters.
>
> *Solution.*

$$R = \frac{\rho}{2\pi L}\left(\log_e \frac{4L}{a} - 1\right)$$
$$= \frac{200,000}{2\pi \times 305}\log_e \frac{4 \times 305}{1.27} = 716\,\Omega$$

13-4. PRACTICAL VALUES OF RESISTIVITY

In the formulas for ground resistance presented in Table 13-1, an assumption is made that the soil is perfectly homogeneous and that the value of the resistivity is of some known value throughout the earth volume surrounding the electrode. This is practically never the case. Many factors influence the value of the resistivity, among which may be listed chemical composition of the soil, moisture content, and temperature. If one were to examine the soil in the vicinity of a ground electrode he might find significant variations in all three of these parameters as he penetrates into the earth.

a. Soil Composition. As shown by Table 13-2, soils of different nature have vastly different values of electrical resistivity, the range being greater than 1,000 to 1. Particularly in glaciated country, where layers of soil have been pushed on top of other soil, stratification may exist. Almost every conceivable combination of strata may be found. Low-resistivity top-

Table 13-1. *Formulas for Calculation of Resistances to Ground**

(Approximate formulas including effects of images. Dimensions must be in centimeters when resistivity is in ohm-centimeters.)

Hemisphere radius a	$R = \dfrac{\rho}{2\pi a}$
One ground rod Length L, radius a	$R = \dfrac{\rho}{2\pi L}\left(\log_e \dfrac{4L}{a} - 1\right)$
Two ground rods $s > L$; spacing s	$R = \dfrac{\rho}{4\pi L}\left(\log_e \dfrac{4L}{a} - 1\right) + \dfrac{\rho}{4\pi s}\left(1 - \dfrac{L^2}{3s^2} + \dfrac{2}{5}\dfrac{L^4}{s^4} \cdots\right)$
Two ground rods $s < L$; spacing s	$R = \dfrac{\rho}{4\pi L}\left(\log_e \dfrac{4L}{a} + \log_e \dfrac{4L}{s} - 2 + \dfrac{s}{2L} - \dfrac{s^2}{16L^2} + \dfrac{s^4}{512L^4} \cdots\right)$
Buried horizontal wire Length $2L$, depth $s/2$	$R = \dfrac{\rho}{4\pi L}\left(\log_e \dfrac{4L}{a} + \log_e \dfrac{4L}{s} - 2 + \dfrac{s}{2L} - \dfrac{s^2}{16L^2} + \dfrac{s^4}{512L^4} \cdots\right)$
Right-angle turn of wire Length of arm L, depth $s/2$	$R = \dfrac{\rho}{4\pi L}\left(\log_e \dfrac{2L}{a} + \log_e \dfrac{2L}{s} - 0.2373 + 0.2146\dfrac{s}{L} + 0.1035\dfrac{s^2}{L^2} - 0.0424\dfrac{s^4}{L^4} \cdots\right)$
Three-point star Length of arm L, depth $s/2$	$R = \dfrac{\rho}{6\pi L}\left(\log_e \dfrac{2L}{a} + \log_e \dfrac{2L}{s} + 1.071 - 0.209\dfrac{s}{L} + 0.238\dfrac{s^2}{L^2} - 0.054\dfrac{s^4}{L^4} \cdots\right)$
Four-point star Length of arm L, depth $s/2$	$R = \dfrac{\rho}{8\pi L}\left(\log_e \dfrac{2L}{a} + \log_e \dfrac{2L}{s} + 2.912 - 1.071\dfrac{s}{L} + 0.645\dfrac{s^2}{L^2} - 0.145\dfrac{s^4}{L^4} \cdots\right)$
Six-point star Length of arm L, depth $s/2$	$R = \dfrac{\rho}{12\pi L}\left(\log_e \dfrac{2L}{a} + \log_e \dfrac{2L}{s} + 6.851 - 3.128\dfrac{s}{L} + 1.758\dfrac{s^2}{L^2} - 0.490\dfrac{s^4}{L^4} \cdots\right)$
Eight-point star Length of arm L, depth $s/2$	$R = \dfrac{\rho}{16\pi L}\left(\log_e \dfrac{2L}{a} + \log_e \dfrac{2L}{s} + 10.98 - 5.51\dfrac{s}{L} + 3.26\dfrac{s^2}{L^2} - 1.17\dfrac{s^4}{L^4} \cdots\right)$
Ring of wire Diameter of ring D, diameter of wire d, depth $s/2$	$R = \dfrac{\rho}{2\pi^2 D}\left(\log_e \dfrac{8D}{d} + \log_e \dfrac{4D}{s}\right)$
Buried horizontal strip Length $2L$, section a by b, depth $s/2$, $b < a/8$	$R = \dfrac{\rho}{4\pi L}\left(\log_e \dfrac{4L}{a} + \dfrac{a^2 - \pi ab}{2(a+b)^2} + \log_e \dfrac{4L}{s} - 1 + \dfrac{s}{2L} - \dfrac{s^2}{16L^2} + \dfrac{s^4}{512L^4} \cdots\right)$
Buried horizontal round plate Radius a, depth $s/2$	$R = \dfrac{\rho}{8a} + \dfrac{\rho}{4\pi s}\left(1 - \dfrac{7}{12}\dfrac{a^2}{s^2} + \dfrac{33}{40}\dfrac{a^4}{s^4} \cdots\right)$
Buried vertical round plate Radius a, depth $s/2$	$R = \dfrac{\rho}{8a} + \dfrac{\rho}{4\pi s}\left(1 + \dfrac{7}{24}\dfrac{a^2}{s^2} + \dfrac{99}{320}\dfrac{a^4}{s^4} \cdots\right)$

*H. B. Dwight, Calculation of Resistance to Ground, *Electrical Engineering*, vol. 55, p. 1319, December 1936.

Table 13-2. *Resistivity of Different Soils*

Soil	Resistance,* $\frac{5}{8}$-in. × 5-ft rods			Resistivity, ohm-cm		
	Av	Min	Max	Av	Min	Max
FILLS (ASHES, CINDERS, BRINE WASTE)	14	3.5	41	2,370	590	7,000
CLAY, SHALE, GUMBO, LOAM	24	2	98	4,060	340	16,300
SAME—WITH VARYING PROPORTIONS OF SAND AND GRAVEL	93	6	800	15,800	1,020	135,000
GRAVEL, SAND, STONES, WITH LITTLE CLAY OR LOAM	554	35	2,700	94,000	59,000	458,000

Bureau of Standards Technical Report 108.

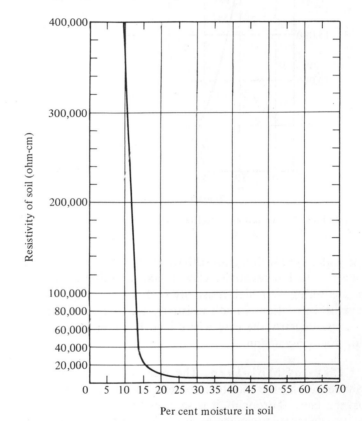

FIG. 13-2. The variation of soil resistivity with moisture content. Red clay soil.

ge
of

(13-8)

value is
maximum
e value of

the ground
. One foot is
stance $r_1 + 1$.
p between two
ius $r_1 + 1$. This

(13-9)

roximately

(13-10)

paragraphs have many
potential if, while standing
that is tied to the ground
out of doors, a man uses a
ystem. A man is subjected to
grounded object while handling
t station. He is, of course, sub
toward or away from the ground
made with a hemispheric electrod

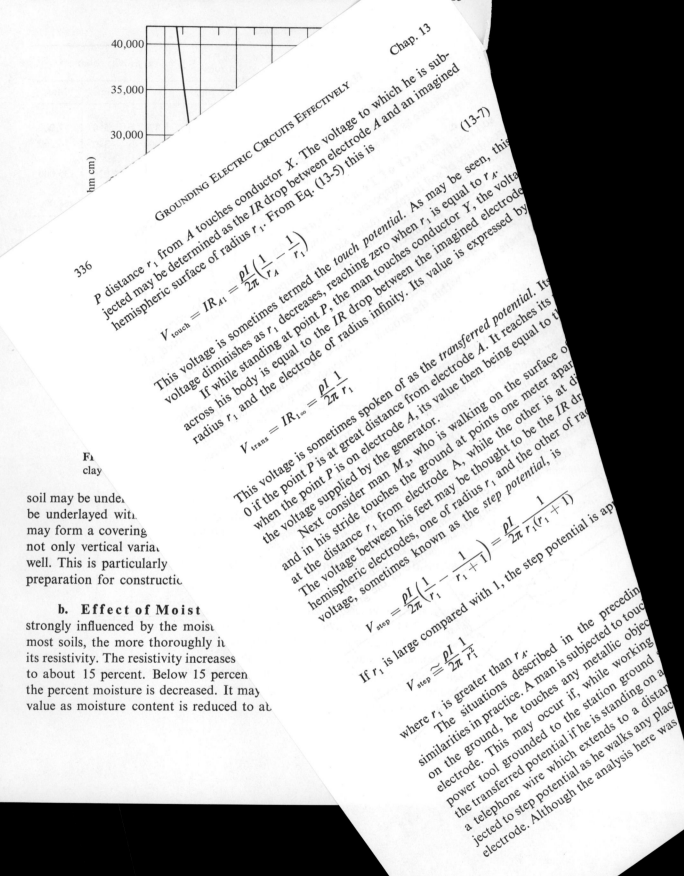

P distance r_1 from A touches conductor X. The voltage to which he is subjected may be determined as the IR drop between electrode A and an imagined hemispheric surface of radius r_1. From Eq. (13-5) this is

$$V_{touch} = IR_{A1} = \frac{\rho I}{2\pi}\left(\frac{1}{r_A} - \frac{1}{r_1}\right) \tag{13-7}$$

This voltage is sometimes termed the *touch potential*. As may be seen, this voltage diminishes as r_1 decreases, reaching zero when r_1 is equal to r_A.

If while standing at point P, the man touches conductor Y, the volta... across his body is equal to the IR drop between the imagined electrode... radius r_1 and the electrode of radius infinity. Its value is expressed by

$$V_{trans} = IR_{1\infty} = \frac{\rho I}{2\pi}\frac{1}{r_1}$$

This voltage is sometimes spoken of as the *transferred potential*. Its ... 0 if the point P is at great distance from electrode A. It reaches its ... when the point P is on electrode A, its value then being equal to t... the voltage supplied by the generator.

Next consider man M_2, who is walking on the surface o... and in his stride touches the ground at points one meter apar... at the distance r_1 from electrode A, while the other is at di... The voltage between his feet may be thought to be the IR dr... hemispheric electrodes, one of radius r_1 and the other of rad... voltage, sometimes known as the *step potential*, is

$$V_{step} = \frac{\rho I}{2\pi}\left(\frac{1}{r_1} - \frac{1}{r_1+1}\right) = \frac{\rho I}{2\pi}\frac{1}{r_1(r_1+1)}$$

If r_1 is large compared with 1, the step potential is ap...

$$V_{step} \cong \frac{\rho I}{2\pi}\frac{1}{r_1^2}$$

where r_1 is greater than r_A.

The situations described in the precedin... similarities in practice. A man is subjected to touc... on the ground, he touches any metallic objec... electrode. This may occur if, while working... power tool grounded to the station ground a... the transferred potential if he is standing on a... a telephone wire which extends to a distan... jected to step potential as he walks any plac... electrode. Although the analysis here was...

soil may be unde...
be underlayed wit...
may form a covering...
not only vertical varia...
well. This is particularly...
preparation for constructi...

b. Effect of Moist...
strongly influenced by the mois...
most soils, the more thoroughly i...
its resistivity. The resistivity increases...
to about 15 percent. Below 15 percen...
the percent moisture is decreased. It may...
value as moisture content is reduced to at...

assumed, a similar analysis made on a driven ground rod or other form of electrode will show the same basic hazards.

Many examples could be cited in which ground-electrode current flow causes stress on the insulation of operating power and communication equipment. Perhaps the most obvious is that which results from the voltage set up between equipment grounded at the station and a telephone line connected to another station some distance away.

13-6. A TWO-ELECTRODE GROUND SYSTEM

A study of a two-electrode ground system aids in showing the behavior characteristics of multipart ground-electrode systems in general. Refer to Fig. 13-5. Two hemispheric electrodes, A and B, of radius r_A and r_B are as-

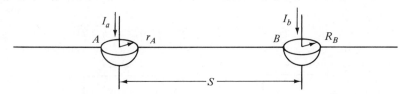

FIG. 13-5. Two hemispheric ground electrodes with radii r_A and r_B and spacing S.

sumed buried flush with the surface of the earth separated by distance S. Current I_a flows into the earth at electrode A and current I_b flows into the earth at electrode B. The return current is from a very large hemispherical dish electrode, not shown.

Assume that the effects of the currents I_a and I_b may be linearly superimposed. This means that the current is assumed to flow radially from each electrode, unaffected by the presence of the other electrode. This is a good approximation provided that S is large compared to r_A and r_B. (If S has a value similar to r_A or r_B more advanced methods of analysis are demanded.)

The voltage on A (measured with respect to remote reference) due to I_a is given by Eq. (13-6) as

$$V_{Aa} = I_a \frac{\rho}{2\pi r_A} \tag{13-11}$$

The voltage on A due to the current I_b is the same as the transferred potential due to current I_b and is given by Eq. (13-8) as

$$V_{Ab} = I_b \frac{\rho}{2\pi S} \tag{13-12}$$

The voltage on A due to I_a and I_b is the sum of the these voltages, or

$$V_A = I_a \frac{\rho}{2\pi r_A} + I_b \frac{\rho}{2\pi S} \tag{13-13}$$

Similarly, the voltage on B due to I_a and I_b is

$$V_B = I_a \frac{\rho}{2\pi S} + I_b \frac{\rho}{2\pi r_B} \tag{13-14}$$

Equations (13-13) and (13-14) may be regarded as the basic equations applying to this two-electrode system. As these are simultaneous equations in I_a, I_b, V_a, and V_b, it follows that if two of these four quantities are specified, the other two are uniquely determined.

Next consider the electric circuit shown in Fig. 13-6, in which the values

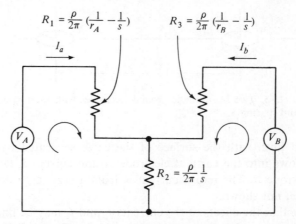

FIG. 13-6. The equivalent electric circuit of two ground electrodes.

of R_1, R_2 and R_3 are specified as shown. If Kirchhoff's voltage equations are written around the two meshes as indicated, simplification of the equations will reproduce Eqs. (13-13) and (13-14). Hence it may be concluded that the three resistances shown in this circuit are equivalent in electrical performance to the two-electrode ground system of Fig. 13-5. From this equivalent circuit diagram, it is possible to determine very quickly the value of resistance which would be measured if the two electrodes of Fig. 13-5 were put in series as shown in Fig. 13-7a. This condition is represented in the equivalent circuit shown in Fig. 13-7b. The resistance measured is

$$R_{\text{series}} = R_1 + R_2 = \frac{\rho}{2\pi}\left(\frac{1}{r_A} + \frac{1}{r_B} - \frac{2}{S}\right) \tag{13-15}$$

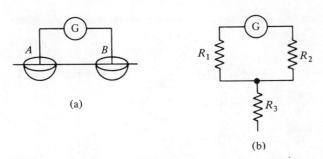

FIG. 13-7. Two ground electrodes in series. (a) The electrodes.
(b) The equivalent circuit.

Similarly, an analysis may be made of the ground resistance measured if the two electrodes of Fig. 13-5 are put in parallel, as shown in Fig. 13-8a. The equivalent circuit of this is shown in Fig. 13-8b. The combined resistance as determined from the equivalent circuit is given by

$$R = \frac{\rho}{2\pi}\left(\frac{1}{2r_K} + \frac{1}{2S}\right) \tag{13-16}$$

assuming that $r_A = r_B = r_K$.

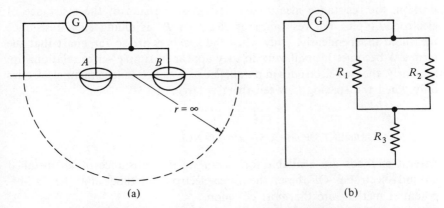

FIG. 13-8. Two ground electrodes in parallel. (a) The electrodes.
(b) The equivalent circuit.

EXAMPLE 13-3. Consider two hemispheric electrodes, each of radius 25 centimeters, with a spacing 400 centimeters in soil of resistivity 8000 ohm centimeters. Calculate

a. R_1, R_2, and R_3.
b. the combined series equivalent resistance.
c. the combined parallel equivalent resistance.

Solution. a. The equivalent circuit of these two electrodes (Fig. 13-6) would have values

$$R_1 = R_2 = \frac{8000}{2\pi}\left(\frac{1}{25} - \frac{1}{400}\right)$$

$$= \frac{8000}{2\pi}(0.040 - 0.0025) = 47.7 \text{ ohms}$$

$$R_3 = \frac{8000}{2\pi}\frac{1}{400} = 3.2 \text{ ohms}$$

b. Then if the two electrodes were connected in series (Fig. 13-7), the resistance of the combination would be

$$R_{\text{series}} = R_1 + R_2 = 47.7 + 47.7 = 95.4 \text{ ohms}$$

c. If the two electrodes were connected in parallel to form a single ground connection (Fig. 13-8), the resistance of the combination would be

$$R = \frac{R_1}{2} + R_3 = 27.0 \text{ ohms}$$

Referring to Eqs. (13-15) and (13-16), it is seen that there are mutual influences between the two ground electrodes. In the case of the series connection, the resistance measured is less than the sum of the resistance of electrode *A* plus the resistance of electrode *B*, assuming each resistance measured independently. Only when the spacings are so far apart that the term $2/S$ becomes insignificant do they appear in simple series relationship. Similarly, the two electrodes in parallel behave as two independent electrodes only when the spacing between them is large.

13-7. MEASUREMENT OF GROUND RESISTANCE

Several methods are available for *measuring* the resistance of an installed ground electrode. Of these, the three-electrode method and the fall-of-potential method are the most common.

    a. **Three-Electrode Method.** Consider a ground electrode *A* whose resistance is to be measured experimentally. This requires that two other electrodes, *B* and *C*, be (temporarily) buried in the ground (Fig. 13-9). The electrodes are shown as hemispheric electrodes. They might be driven ground rods or of other shape. The mathematics of the hemispheric electrodes has been presented and can be readily used here. It is necessary that the electrode spacings are large compared to their radii; otherwise errors are introduced. A bridge or other measuring circuit connected with one terminal

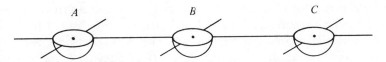

FIG. 13-9. Three ground electrodes provide a system for measuring ground electrode resistance.

on electrode A and one on electrode B measures R_X as

$$R_X = R_A + R_B \qquad (13\text{-}17)$$

Similarly, R_Y is measured between electrodes B and C, and R_Z between electrodes C and A with the result

$$R_Y = R_B + R_C \qquad (13\text{-}18)$$

$$R_Z = R_C + R_A \qquad (13\text{-}19)$$

Solving these three equations simultaneously gives the value for R_A as

$$R_A = \frac{R_X + R_Z - R_Y}{2} \qquad (13\text{-}20)$$

This method of measurement gives best results if the three resistors are of approximately the same resistance magnitude. If R_B or R_C is large compared to R_A, the calculated value of R_A is subject to considerable experimental error.

b. Fall-of-Potential Method. The resistance of a ground connection may also be measured by the *fall-of-potential* method presented in Fig. 13-10. Again A is the electrode under test whose radius is r_A. Current

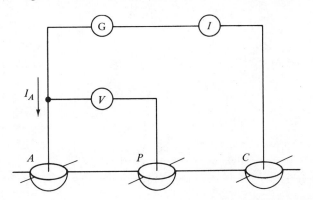

FIG. 13-10. The fall-of-potential method of measuring ground resistance.

from the generator G is sent into the earth through electrode A and returned from a current electrode C. The voltage V is measured between the electrode A and a potential electrode P. The three electrodes need *not* be in a straight line, as shown in Fig. 13-10. If all electrodes are very widely separated, the voltmeter V will measure substantially the same voltage as if connected from electrode A to a remote reference point. The resistance then is

$$R_A = \frac{V}{I} \qquad\qquad (13\text{-}21)$$

In a practical situation, it is sometimes very difficult to provide physical separations that are large compared to the radius of the electrode under test. When measurements are made under these adverse conditions, the voltage of electrode A is determined not only by the current into A but also by the current flow field of electrode C. This behavior is similar to that shown by Eq. (13-13). Similarly, a voltage appears on point P due to the two current-flow fields. Therefore, P cannot be regarded as a reference point at infinity. Under such conditions, the measured value of resistance may be in error.

There are, however, certain electrode arrangements which (theory indicates) yield the correct value for the resistance of electrode A. One arrangement is shown in Fig. 13-11. With electrodes A and C located as shown, a correct reading of the resistance of electrode A is obtained with the electrode P at either X, Y, or Z. Considering the distance from A to C to be 100 percent, the point X is located 62 percent of the distance from A. The point Y is located 162 percent of the distance from A to C, the point Z is located 100 percent of the distance from A and 50 percent of the distance from C.

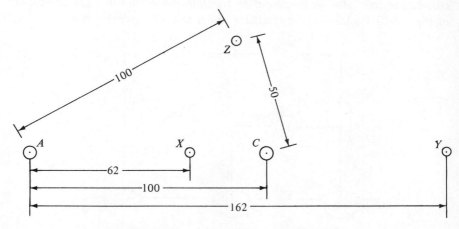

FIG. 13-11. Electrode arrangements which minimize error in the fall-of-potential method of measurement.

Although the measuring arrangement just described is theoretically correct for the assumption of uniform soil resistivity, it may be somewhat in error under the nonuniform conditions that always prevail.

   c. Precautions in Measurements. Certain precautions are necessary in using either the three-electrode method or the fall-of-potential method in measuring ground resistance. The ground connection to be measured must be insulated from connections which may tie it to other grounds remotely located. For example, overhead ground wires of transmission lines sometimes are connected to station structures. The ground-wire connection at the structure should be removed or insulated. Care should be taken to make certain that all such connections are removed.

   The leads that run from the measuring instruments to the three ground electrodes provide situations similar to those with which touch and transferred potentials were described in Section 13-5. In case a fault occurs on the power system of such nature as to cause high current flow through the ground under test, dangerous values of touch and transferred potential may be produced. A person making tests must protect himself against the possibility of a dangerous voltage appearing on his test leads.

13-8. MEASUREMENT OF SOIL RESISTIVITY

Soil resistivity may be measured by several different methods, two of which are described. As all the formulas for ground resistance given in Sec. 13-3 have, as a common factor, the resistivity of the soil, it is apparent that a knowledge of soil resistivity is helpful to the designer of a ground-electrode system to be installed at a particular location.

   a. Four-Electrode Method of Measuring Resistivity. A commonly used method of measuring earth resistivity is described with reference to Fig. 13-12. Four electrodes, *A*, *B*, *C*, and *D*, are

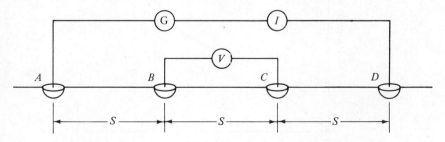

FIG. 13-12. The four-electrode method of measuring earth resistivity.

placed in the ground in a straight line with spacing S. In the diagram these electrodes are shown as being hemispheric in shape, although this is not a requirement. Short driven rods serve equally well and are more convenient to obtain and use. A current I is circulated through the ground between electrodes A and D. The voltage, V, between B and C is measured. Then

$$\rho = \frac{V}{I} 2\pi S \tag{13-22}$$

The spacing of the probes influences the depth of penetration of the principal part of the current between the electrodes A and D. For this reason, wide spacing S between the electrodes results in a measurement of the resistivity deep within the ground. It is sometimes assumed that the value of resistivity measured is approximately equal to the average value of the resistivity of the soil down to a depth S.

b. Driven-Ground-Rod Method. The resistivity of soil at varying depths may be determined quite accurately by driving a ground rod into the soil and making a measurement of its resistance, periodically, during the driving process. A curve is then plotted showing the rod conductance G plotted against *depth L* (Fig. 13-13). In Sec. 13-3 the formula for the resistance of a driven ground rod was given as

$$R = \frac{\rho}{2\pi L}\left(\log_\epsilon \frac{4L}{a} - 1\right) \tag{13-23}$$

This expression may be inverted to give the conductance of a ground rod:

$$G = L\frac{2\pi}{\rho}\,\frac{1}{\log_\epsilon (4L/a) - 1} \tag{13-24}$$

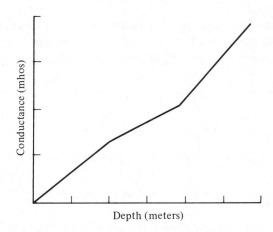

FIG. 13-13. Ground-rod conductance vs. depth.

Equation (13-24) shows the conductance of a driven rod to be the product of three terms, one is the length L, one is $2\pi/\rho$, and one is a logarithmic expression involving the ratio of rod length to rod radius. It may be observed that the term under the horizontal line is a very slowly changing function of L. For a rod of 1/2 inch radius, this term changes only from 4.2 to 6.5 as L is increased from 2 feet to 20 feet. If it is assumed to be constant, at a value 5.0, then an approximate expression for G is

$$G = \frac{2\pi L}{5\rho} \tag{13-25}$$

or

$$G = 0.4\pi L\sigma \tag{13-26}$$

where σ is the soil conductivity, the reciprocal of ρ.

Differentiating Eq. (13-26) with respect to L results in

$$\frac{dG}{dL} = 0.4\pi\sigma \tag{13-27}$$

or

$$\sigma = \frac{10}{4\pi}\frac{dG}{dL} \tag{13-28}$$

The value of σ derived from this relationship is the value applying to the soil at that depth for which dG/dL is calculated. L must be expressed in centimeters if σ is to be in mhos per centimeter. Then $\rho = 1/\sigma$ is in ohm-centimeters.

EXAMPLE 13-4. Suppose that a rod of $\frac{1}{2}$-inch radius is driven into the ground and resistance measurements are taken at various depths. At 8 feet (244 centimeters) the resistance is 500 ohms (0.0020 mho) and at 12 feet (366 centimeters) the resistance is 400 ohms (0.0025 mho). A plot of the conductance in this region is as shown in Fig. 13-14. The slope of the conductance line is

$$\text{slope} = \frac{\Delta G}{\Delta L} = \frac{0.005}{122} = 0.000041 \text{ mho/cm}$$

The conductivity in the region from 8 to 12 feet deep is then, by Eq. (13-28),

$$\sigma = \frac{10}{4\pi} \times 0.000041 = 0.000033 \text{ mho/cm}$$

and the resistivity is

$$\rho = \frac{1}{\sigma} = 30,400 \text{ ohm-cm}$$

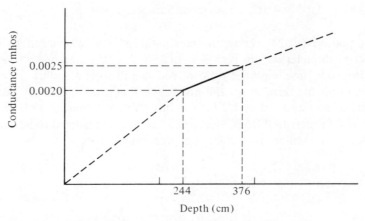

FIG. 13-14. An experimentally determined conductance-depth relationship of a driven ground rod.

In making ground-resistivity measurements near a ground electrode of an operating power system, safety precautions must be taken similar to those mentioned for the measurement of resistance in Sec. 13-7. In case of a heavy current flow through the power-system ground, voltages of dangerous magnitude may appear between the electrodes used for resistivity measurements.

In making measurements of resistance and resistivity it is necessary to *avoid* using direct current instruments. The conduction of current in the ground is by a process very similar to that of the conduction of the current in a liquid. With a unidirectional flow of current, a process known as polarization causes gas to accumulate near one of the electrodes, with an accompanying increase in the apparent resistance of that electrode. The use of alternating current for resistance and resistivity measurements entirely eliminates this effect. Self-contained ground-resistance measuring instruments are available which provide an alternating voltage source and the associated instrumentation.

13-9. LIMITATIONS ON GROUND-ELECTRODE DESIGN

In designing and constructing a ground-electrode system it may be hoped to meet certain specifications on the resistance of the finished system. From first considerations it might be thought that by putting enough electrode material in the earth, the ground resistance of the complete system could be made as low as desired. However, this is not the case if the land area for the construction of the electrode system is limited. This subject may be discussed more readily by referring to the conductance of the electrode system rather than its resistance. To attempt to reduce its resistance to a specified value is equivalent to attempting to raise its conductance to some value.

The limitations on the design of a ground-electrode system on a restricted area will be illustrated by reference to a system consisting of a number of driven ground rods. Suppose that a number of $\frac{3}{4}$-inch diameter, 10-foot rods are to be driven in a limited area to form a ground connection. The conductance of the system increases each time another rod is driven. However, the conductance of two rods is slightly less than twice the conductance of one rod, and as other rods are added the conductance increases but at a diminishing rate. Fig. 13-15 shows the ratio of the conductance of a group of rods to the conductance of one rod plotted against the number of rods. Separate curves are shown for different available ground surface areas. It may be noted that for an area of 500 square feet, 10 rods have about 4.7 times the conductance of a single rod. An infinite number of rods in that area would have a conductance only 4.9 times the conductance of one rod. This limitation may seem more reasonable when it is considered that an infinite number of 10-foot rods on the 500-square foot area would afford a ground electrode equivalent to that of a metal box 10 feet deep filling the entire 500 square foot area. This box might be compared with a hemispheric

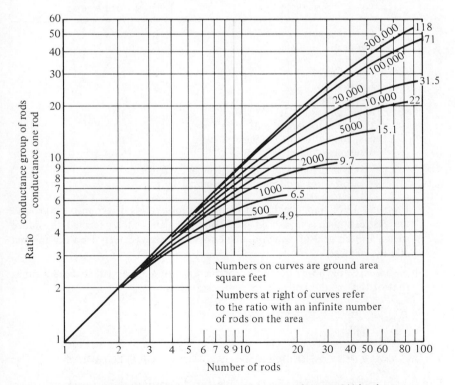

FIG. 13-15. Comparison of the conductance of many 3/4-in. by 10-ft ground rods spread over an area with the conductance of a single ground rod.

electrode of the same surface area, the resistance (or conductance) of which was discussed in Section 13-2.

In some cases, ground electrode systems are made of grids of buried horizontal wires. With such an arrangement, increasing the number of meshes in the grid increases its conductance. However, its conductance can never be increased beyond that of a buried plate of the same outside dimensions.

In some instances, a grid of buried horizontal wires is used in combination with driven ground rods. Table 13-3 shows a comparison of the conductance of different forms of ground-electrode systems, each restricted to a ground area 100 feet square. Attention must be given to the inherent limitations of the conductance of ground-electrode systems to avoid the expenditure of unnecessary time and materials in an effort to obtain a conductance that is, in fact, unattainable.

Table 13-3. *Comparison of the Conductance of a Single Ground Rod with the Conductance of Other Ground-Electrode Systems Land Area 100 feet by 100 feet*

Structure*	Relative Conductance
1 GROUND ROD	1.0
4 RODS	3.6
36 RODS	8.75
1-MESH GRID	7.0
4-MESH GRID	7.6
25-MESH GRID	8.7
PLATE, 100 by 100 ft	10.0
25-MESH GRID plus 36 rods	10.0
BOX, 100 by 100 ft, 20 ft deep	12.1
HEMISPHERE, diameter 100 ft	16.1

*Ground rods are 1-in. diameter by 20 ft deep
Grid wires are 1-in. diameter buried $1\frac{1}{2}$ ft

EXAMPLE 13-5. Suppose that a 1-inch by 20-foot rod is driven into the ground and its resistance is measured as 250 ohms. Calculate the expected resistance of a 25-mesh grid of 1-inch-diameter wire buried $1\frac{1}{2}$ feet deep on an area 100 by 100 feet.

Solution. Table 13-3 shows the conductance of such a grid to be 8.7 times that of the ground rod. Hence its resistance is

$$R_{\text{grid}} = \frac{R_{\text{rod}}}{8.7} = \frac{250}{8.7} = 29 \text{ ohms}$$

13-10. ANALYTICAL AND EXPERIMENTAL STUDIES OF ELECTRODES

Digital-computer methods are available for determining the resistance of a ground-electrode system of any configuration. The mathematical methods of analysis and the computer programming is beyond the scope of the present discussion. Practically all these methods of calculation are based on the assumption of uniform soil resistivity, a condition that is almost never found. For this reason the great precision of computer methods is of little help.

Ground-electrode-system designs may also be studied by means of scale models. The models may be constructed on a sheet of $\frac{3}{8}$-inch plywood, to support it at the surface of the water of a lake or swimming pool. The measurement of the resistance of the model is made in a fashion exactly like that described in Sec. 13-7 and Fig. 13-10.

Whenever a model is used it is necessary to determine the relations by which it is scaled. Consider the hemispheric electrode discussed in Section 13-2. Doubling the radius of such an electrode multiplies its volume by 8 and its surface area by 4. Reference to Eq. (13-6) shows that its resistance is divided by 2 or its conductance is doubled. Similarly, reference to the expression for the resistance of a driven ground rod Eq. (13-23) shows that if all dimensions of the rod are multiplied by 2, the conductance is doubled.

It may be shown that this form of relationship exists for all possible ground-electrode systems. If all the linear dimensions of the parts of a system are multiplied by the factor F, the conductance is multiplied by F. This relationship shows the simplicity of the study of electrode systems by means of scale models.

PROBLEMS

13-1. List, from your own observations, items of electrical equipment that are grounded.

13-2. Determine the resistance of a $\frac{1}{2}$-in. by 8-ft ground rod driven into
   a. average clay soil.
   b. high-resistivity sand.

13-3. Compare the resistance of a hemisphere electrode of diameter 50 cm with the resistance of a 50-cm-diameter ring of wire. The wire is 1 cm in diameter buried 20 cm deep. Specify the soil resistivity considered.

13-4. A ground-electrode system in red clay soil of 20 percent moisture content measures 35 ohms. What resistance might be expected if in a dry season, the moisture content dropped to 10 percent?

13-5. A ground-electrode system in clay soil measures 50 ohms when the temperature is 40°F. What resistance might be expected at 10°F?

13-6. A 25-cm-radius ground electrode is buried in soil of resistivity 8000 ohm-cm. A current of 500 amperes flows to ground through the electrode. A man is standing 6 meters from the electrode. Determine the voltage to which he will be subjected if he
   a. touches a wire that connects to the electrode.
   b. touches a wire that connects to a remote location.
   c. in walking, touches the ground at two points 1 meter apart.

13-7.   Three ground electrodes, *A*, *B*, and *C*, are widely spaced relative to their diameters. Resistances are measured by the three-electrode method with the following results:

A and *B*	370 ohms
B and *C*	750 ohms
C and *A*	420 ohms

Determine the resistance of each electrode.

13-8.   Refer to Problem 13-7. Suppose that in making the *A* and *B* measurement, the value recorded was 5 percent too high. What then is the calculated resistance of electrode *A*? What percent error results?

13-9.   The resistance of a ground-electrode system is made by the fall-of-potential method. The voltage of the source is 60 volts, the voltage between the subject electrode and the potential electrode is 21 volts, and the current is 5.1 amperes. What is the value of the ground resistance?

13-10. Ground resistivity is measured by the four-electrode method (Fig. 13-12), with electrodes spaced 12 feet apart. Observations show 32 volts and 0.15 amperes. Determine the soil resistivity in ohm-centimeters.

13-11. Refer to Problem 13-10. Suppose that electrodes *B* and *C* each had a resistance of 5000 ohms and the voltmeter had a resistance of 15,000 ohms. What correction in the results would be indicated? Is this same type of error possible in measuring resistance by the fall-of-potential method?

13-12. One $\frac{3}{4}$-in. by 10-foot ground rod has a measured ground resistance of 480 ohms. What will be the expected resistance of 20 rods spread over an area 50 by 100 feet?

13-13. Refer to Problem 13-12. What will be the expected resistance of a great number of rods spread over an area 100 by 100 feet?

13-14. Assume that the ground resistance is 2000 ohm-cms. Determine the resistance of

a. a 1-inch by 20-foot ground rod.
b. a 25-mesh grid with 36 rods, spread over an area 100 by 100 feet.
c. a hemisphere 100 feet in diameter.

13-15. A 20-foot, 1-inch-diameter rod is driven into the ground and resistance measurements taken as follows:

Depth, ft	Resistance, ohms
5	4000
10	2000
15	800
20	300

Determine the approximate soil resistivity at 17 feet.

Tables For The Determination of Circuit Constants

Table A-1. *Resistivity and Temperature Coefficient of Conductor Materials*

Material	Resistivity (ρ) at 20°C Micro-ohm cm	Ohms cir mil per ft	Temp. Coefficient (α) at 20°C
ALUMINUM	2.83	17.0	0.0039
BRASS	6.4–8.4	38–51	0.002
BRONZE	13–18	78–108	0.0005
COPPER			
Hard drawn	1.77	10.62	0.00382
Annealed	1.72	10.37	0.00393
IRON	10	60	0.0050
SILVER	1.59	9.6	0.0038
SODIUM	4.3	26	0.0044
STEEL	12–88	72–530	0.001–0.005

Table A-2. *Characteristics of COPPER Conductors, Hard Drawn, 97.3 Percent*

Size of Conductor		Number of Strands	Diameter of Individual Strands Inches	Outside Diameter Inches	Breaking Strength Pounds	Weight Pounds per Mile	Approx. Current Carrying Capacity* Amps	Geometric Mean Radius at 60 Hz Feet
Circular Mils	A.W.G. or B. & S.							
1 000 000	....	37	0.1644	1.151	43 830	16 300	1 300	0.0368
900 000	....	37	0.1560	1.092	39 510	14 670	1 220	0.0349
800 000	....	37	0.1470	1.029	35 120	13 040	1 130	0.0329
750 000	....	37	0.1424	0.997	33 400	12 230	1 090	0.0319
700 000	....	37	0.1375	0.963	31 170	11 410	1 040	0.0308
600 000	....	37	0.1273	0.891	27 020	9 781	940	0.0285
500 000	....	37	0.1162	0.814	22 510	8 151	840	0.0260
500 000	....	19	0.1622	0.811	21 590	8 151	840	0.0256
450 000	....	19	0.1539	0.770	19 750	7 336	780	0.0243
400 000	....	19	0.1451	0.726	17 560	6 521	730	0.0229
350 000	....	19	0.1357	0.679	15 590	5 706	670	0.0214
350 000	....	12	0.1708	0.710	15 140	5 706	670	0.0225
300 000	....	19	0.1257	0.629	13 510	4 891	610	0.01987
300 000	....	12	0.1581	0.657	13 170	4 891	610	0.0208
250 000	....	19	0.1147	0.574	11 360	4 076	540	0.01813
250 000	....	12	0.1443	0.600	11 130	4 076	540	0.01902
211 600	4/0	19	0.1055	0.528	9 617	3 450	480	0.01668
211 600	4/0	12	0.1328	0.552	9 483	3 450	490	0.01750
211 600	4/0	7	0.1739	0.522	9 154	3 450	480	0.01579
167 800	3/0	12	0.1183	0.492	7 556	2 736	420	0.01159
167 800	3/0	7	0.1548	0.464	7 366	2 736	420	0.01404
133 100	2/0	7	0.1379	0.414	5 926	2 170	360	0.01252
105 000	1/0	7	0.1228	0.368	4 752	1 720	310	0.01113
83 690	1	7	0.1093	0.328	3 804	1 364	270	0.00992
83 690	1	3	0.1670	0.360	3 620	1 351	270	0.01016
66 370	2	7	0.0974	0.292	3 045	1 082	230	0.00883
66 370	2	3	0.1487	0.320	2 913	1 071	240	0.00903
66 370	2	1	....	0.258	3 003	1 061	220	0.00836
52 630	3	7	0.0867	0.260	2 433	858	200	0.00787
52 630	3	3	0.1325	0.285	2 359	850	200	0.00805
52 630	3	1	....	0.229	2 439	841	190	0.00745
41 740	4	3	0.1180	0.254	1 879	674	180	0.00717
41 740	4	1	....	0.204	1 970	667	170	0.00663
33 100	5	3	0.1050	0.226	1 505	534	150	0.00638
33 100	5	1	....	0.1819	1 591	529	140	0.00590
26 250	6	3	0.0935	0.201	1 205	424	130	0.00568
26 250	6	1	....	0.1620	1 280	420	120	0.00526
20 820	7	1	....	0.1443	1 030	333	110	0.00468
16 510	8	1	....	0.1285	826	264	90	0.00417

*For conductor at 75°C., air at 25°C., wind 1.4 miles per hour (2 ft/sec), frequency = 60 Hz

Conductivity

r_a Resistance Ohms per Mile per Conductor						X_a Inductive Reactance Ohms per mile per Conductor		X'_a Capacitive Reactance Megohm-miles per Conductor	
25°C.(77°F.)			50°C.(122°F.)						
d-c	50 Hz	60 Hz	d-c	50 Hz	60 Hz	50 Hz	60 Hz	50 Hz	60 Hz
0.0585	0.0620	0.0634	0.0640	0.0672	0.0685	0.333	0.400	0.1081	0.0901
0.0650	0.0682	0.0695	0.0711	0.0740	0.0752	0.339	0.406	0.1100	0.0916
0.0731	0.0760	0.0772	0.0800	0.0826	0.0847	0.344	0.413	0.1121	0.0934
0.0780	0.0807	0.0818	0.0853	0.0878	0.0888	0.348	0.417	0.1132	0.0943
0.0836	0.0861	0.0871	0.0914	0.0937	0.0947	0.352	0.422	0.1145	0.0954
0.0975	0.0997	0.1006	0.1066	0.1086	0.1095	0.360	0.432	0.1173	0.0977
0.1170	0.1188	0.1196	0.1280	0.1296	0.1303	0.369	0.443	0.1205	0.1004
0.1170	0.1188	0.1196	0.1280	0.1296	0.1303	0.371	0.445	0.1206	0.1005
0.1300	0.1316	0.1323	0.1422	0.1437	0.1443	0.376	0.451	0.1224	0.1020
0.1462	0.1477	0.1484	0.1600	0.1613	0.1619	0.382	0.458	0.1245	0.1038
0.1671	0.1684	0.1690	0.1828	0.1840	0.1845	0.389	0.466	0.1269	0.1058
0.1671	0.1684	0.1690	0.1828	0.1840	0.1845	0.384	0.460	0.1253	0.1044
0.1950	0.1961	0.1966	0.213	0.214	0.215	0.396	0.476	0.1296	0.1080
0.1950	0.1961	0.1966	0.213	0.214	0.215	0.392	0.470	0.1281	0.1068
0.234	0.235	0.235	0.256	0.257	0.257	0.406	0.487	0.1329	0.1108
0.234	0.235	0.235	0.256	0.257	0.257	0.401	0.481	0.1313	0.1094
0.276	0.277	0.278	0.302	0.303	0.303	0.414	0.497	0.1359	0.1132
0.276	0.277	0.278	0.302	0.303	0.303	0.409	0.491	0.1343	0.1119
0.276	0.277	0.278	0.302	0.303	0.303	0.420	0.503	0.1363	0.1136
0.349	0.349	0.350	0.381	0.382	0.382	0.421	0.505	0.1384	0.1153
0.349	0.349	0.350	0.381	0.382	0.382	0.431	0.518	0.1405	0.1171
0.440	0.440	0.440	0.481	0.481	0.481	0.443	0.532	0.1445	0.1205
0.555	0.555	0.555	0.606	0.607	0.607	0.455	0.546	0.1488	0.1240
0.699	0.699	0.699	0.765			0.467	0.560	0.1528	0.1274
0.692	0.692	0.692	0.757			0.464	0.557	0.1495	0.1246
0.881	0.882	0.882	0.964			0.478	0.574	0.1570	0.1308
0.873			0.955			0.476	0.571	0.1537	0.1281
0.864			0.945			0.484	0.581	0.1641	0.1345
1.112			1.216			0.490	0.588	0.1611	0.1343
1.101			1.204			0.488	0.585	0.1578	0.1315
1.090	Same as d-c		1.192	Same as d-c		0.496	0.595	0.1656	0.1380
1.388			1.518			0.499	0.599	0.1619	0.1349
1.374			1.503			0.507	0.609	0.1697	0.1415
1.750			1.914			0.511	0.613	0.1661	0.1384
1.733			1.895			0.519	0.623	0.1738	0.1449
2.21			2.41			0.523	0.628	0.1703	0.1419
2.18			2.39			0.531	0.637	0.1779	0.1483
2.75			3.01			0.542	0.651	0.1821	0.1517
3.47			3.80			0.554	0.665	0.1862	0.1552

Table A-3. *Characteristics of Aluminum Cable Steel Reinforced*

Circular Mils or A.W.G. Aluminum	Aluminum			Steel		Outside Diameter Inches	Copper Equivalent* Circular Mils or A.W.G.	Ultimate Strength Pounds	Weight Pounds per Mile	Geometric Mean Radius at 60 Hz Feet	Approx. Current Carrying Capacity† Amps
	Strands	Layers	Strand Dia. Inches	Strand	Strand Dia. Inches						
1 590 000	54	3	0.1716	19	0.1030	1.545	1 000 000	56 000	10 777	0.0520	1 389
1 510 500	54	3	0.1673	19	0.1004	1.506	950 000	53 200	10 237	0.0507	1 340
1 431 000	54	3	0.1628	19	0.0977	1.465	900 000	50 400	9 699	0.0493	1 300
1 351 000	54	3	0.1582	19	0.0949	1.424	850 000	47 600	9 160	0.0479	1 250
1 272 000	54	3	0.1535	19	0.0921	1.382	800 000	44 800	8 621	0.0465	1 200
1 192 500	54	3	0.1486	19	0.0892	1.338	750 000	41 100	8 082	0.0450	1 160
1 113 000	54	3	0.1436	19	0.0862	1.293	700 000	40 200	7 544	0.0435	1 110
1 033 500	54	3	0.1384	7	0.1384	12.46	650 000	37 100	7 019	0.0420	1 060
954 000	54	3	0.1329	7	0.1329	1.196	600 000	34 200	6 479	0.0403	1 010
900 000	54	3	0.1291	7	0.1291	1.162	566 000	32 300	6 112	0.0391	970
874 000	54	3	0.1273	7	0.1273	1.146	550 000	31 400	5 940	0.0386	950
795 000	54	3	0.1214	7	0.1214	1.093	500 000	28 500	5 399	0.0368	900
795 000	26	2	0.1749	7	0.1360	1.108	500 000	31 200	5 770	0.0375	900
795 000	54	3	0.1628	19	0.0977	1.140	500 000	38 400	6 517	0.0393	910
715 500	54	3	0.1151	7	0.1151	1.036	450 000	26 300	4 859	0.0349	830
715 500	26	2	0.1659	7	0.1290	1.051	450 000	28 100	5 193	0.0355	840
715 500	30	2	0.1544	19	0.0926	1.081	450 000	34 600	5 865	0.0372	840
666 600	54	3	0.1111	7	0.1111	1.000	419 000	24 500	4 527	0.0337	800
636 000	54	3	0.1085	7	0.1085	0.977	400 000	23 600	4 319	0.0329	770
636 000	26	2	0.1564	7	0.1216	0.990	400 000	25 000	4 616	0.0335	780
636 000	30	2	0.1456	19	0.0874	1.019	400 000	31 500	5 213	0.0351	780
605 000	54	3	0.1059	7	0.1059	0.953	380 500	22 500	4 109	0.0321	750
605 000	26	2	0.1525	7	0.1186	0.966	380 500	24 100	4 391	0.0327	760
556 500	26	2	0.1463	7	0.1138	0.927	350 000	22 400	4 039	0.0313	730
556 500	30	2	0.1362	7	0.1362	0.953	350 000	27 200	4 588	0.0328	730
500 000	30	2	0.1291	7	0.1291	0.904	314 500	24 400	4 122	0.0311	690
477 000	26	2	0.1355	7	0.1054	0.858	300 000	19 430	3 462	0.0290	670
477 000	30	2	0.1261	7	0.1261	0.883	300 000	23 300	3 933	0.0304	670
397 500	26	2	0.1236	7	0.0961	0.783	250 000	16 190	2 885	0.0265	590
397 500	30	2	0.1151	7	0.1151	0.806	250 000	19 980	3 277	0.0278	600
336 400	26	2	0.1138	7	0.0885	0.721	4/0	14 050	2 442	0.0244	530
336 400	30	2	0.1059	7	0.1059	0.741	4/0	17 040	2 774	0.0255	530
300 000	26	2	0.1074	7	0.0835	0.680	188 700	12 650	2 178	0.0230	490
300 000	30	2	0.1000	7	0.1000	0.700	188 700	15 430	2 473	0.0241	500
266 800	26	2	0.1013	7	0.0788	0.642	3/0	11 250	1 936	0.0217	460
										For Current Approx. 75% Capacity‡	
266 800	6	1	0.2109	7	0.0703	0.633	3/0	9 645	1 802	0.00684	460
4/0	6	1	0.1878	1	0.1878	0.563	2/0	8 420	1 542	0.00814	340
3/0	6	1	0.1672	1	0.1672	0.502	1/0	6 675	1 223	0.00600	300
2/0	6	1	0.1490	1	0.1490	0.447	1	5 345	970	0.00510	270
1/0	6	1	0.1327	1	0.1327	0.398	2	4 280	769	0.00446	230
1	6	1	0.1182	1	0.1182	0.335	3	3 480	610	0.00418	200
2	6	1	0.1052	1	0.1052	0.316	4	2 790	484	0.00418	180
2	7	1	0.0974	1	0.1299	0.325	4	3 525	566	0.00504	180
3	6	1	0.0936	1	0.0937	0.281	5	2 250	384	0.00430	160
4	6	1	0.0834	1	0.0834	0.250	6	1 830	304	0.00437	140
4	7	1	0.0772	1	0.1029	0.257	6	2 288	356	0.00452	140
5	6	1	0.0743	1	0.0743	0.223	7	1 460	241	0.00416	241
6	6	1	0.0661	1	0.0661	0.198	8	1 170	191	0.00394	100

*Based on copper 97 percent, aluminum 61 percent conductivity.

†For conductor at 75°C., air at 25°C., wind 1.4 miles per hour (2 ft/sec), frequency = 60 Hz.

‡"Current Approx. 75% Capacity" is 75% of the "Approx. Current Carrying Capacity in Amps." and is approximately the current which will produce 50°C. conductor temp. (25°C. rise) with 25°C. air temp., wind 1.4 miles per hour.

Top Table

| r_a Resistance Ohms per Mile per Conductor | | | | | | X_a Inductive Reactance Ohms per Mile per Conductor | | X'_a Capacitive Reactance Megohm-miles per Conductor | |
| 25°C.(77°F.) Small Currents | | | 50°C. (122°F.) Current Approx. 75% Capacity‡ | | | | | | |
d-c	50 Hz	60 Hz	d-c	50 Hz	60 Hz	50 Hz	60 Hz	50 Hz	60 Hz
0.0587	0.0590	0.0591	0.0466	0.0675	0.0684	0.299	0.359	0.0977	0.0814
0.0618	0.0621	0.0622	0.0680	0.0710	0.0720	0.302	0.362	0.0986	0.0821
0.0652	0.0655	0.0656	0.0718	0.0749	0.0760	0.304	0.365	0.0996	0.0830
0.0691	0.0694	0.0695	0.0761	0.0792	0.0803	0.307	0.369	0.1006	0.0838
0.0734	0.0737	0.0738	0.0808	0.0840	0.0851	0.310	0.372	0.1016	0.0847
0.0783	0.0786	0.0788	0.0862	0.0894	0.0906	0.314	0.376	0.1028	0.0857
0.0839	0.0842	0.0844	0.0924	0.0957	0.0969	0.317	0.380	0.1040	0.0867
0.0903	0.0907	0.0909	0.0994	0.1025	0.1035	0.321	0.385	0.1053	0.0878
0.0979	0.0981	0.0982	0.1078	0.1180	0.1128	0.325	0.390	0.1068	0.0890
0.104	0.104	0.104	0.1145	0.1175	0.1185	0.328	0.393	0.1078	0.0898
0.107	0.107	0.108	0.1178	0.1218	0.1228	0.329	0.395	0.1083	0.0903
0.117	0.118	0.119	0.1288	0.1358	0.1378	0.334	0.401	0.1100	0.0917
0.117	0.117	0.117	0.1288	0.1288	0.1288	0.332	0.399	0.1095	0.0912
0.117	0.117	0.117	0.1288	0.1288	0.1288	0.327	0.393	0.1085	0.0904
0.131	0.131	0.132	0.1442	0.1472	0.1482	0.339	0.407	0.1119	0.0932
0.131	0.131	0.131	0.1442	0.1442	0.1442	0.337	0.405	0.1114	0.0928
0.131	0.131	0.131	0.1442	0.1442	0.1442	0.333	0.399	0.1104	0.0920
0.140	0.141	0.141	0.1541	0.1591	0.1601	0.343	0.412	0.1132	0.0943
0.147/	0.148	0.148	0.1618	0.1678	0.1688	0.345	0.414	0.1140	0.0950
0.147	0.147	0.147	0.1618	0.1618	0.1614	0.344	0.412	0.1135	0.0946
0.147	0.147	0.147	0.1618	0.1618	0.1618	0.339	0.406	0.1125	0.0937
0.154	0.155	0.155	0.1695	0.1755	0.1775	0.343	0.417	0.1149	0.0957
0.154	0.154	0.154	0.1700	0.1720	0.1720	0.346	0.415	0.1144	0.0953
0.168	0.168	0.168	0.1849	0.1859	0.1859	0.350	0.420	0.1159	0.0965
0.168	0.168	0.168	0.1849	0.1859	0.1859	0.346	0.415	0.1149	0.0957
0.187	0.187	0.187	0.206			0.351	0.421	0.1167	0.0973
0.196	0.196	0.196	0.216			0.358	0.430	0.1186	0.0988
0.196	0.196	0.196	0.216			0.353	0.424	0.1176	0.0980
0.235			0.259			0.367	0.441	0.1219	0.1015
0.235	Same as d-c		0.259	Same as d-c		0.362	0.435	0.1208	0.1006
0.278			0.306			0.376	0.451	0.1248	0.1039
0.278			0.306			0.371	0.445	0.1238	0.1032
0.311			0.342			0.382	0.458	0.1269	0.1057
0.311			0.342			0.377	0.452	0.1258	0.1049
0.350			0.385			0.387	0.465	0.1289	0.1074

Bottom Table (Single Layer Conductors)

| r_a Resistance Ohms per Mile per Conductor | | | | | | X_a Inductive Reactance — Single Layer Conductors | | | | X'_a Capacitive Reactance Megohm-miles per Conductor | |
| 25°C.(77°F.) Small Currents | | | 50°C. (122°F.) Current Approx. 75% Capacity‡ | | | Small Currents | | Current Approx. 75% Capacity‡ | | | |
d-c	50 Hz	60 Hz	d-c	50 Hz	60 Hz	50 Hz	60 Hz	50 Hz	60 Hz	50 Hz	60 Hz
0.351	0.351	0.352	0.386	0.510	0.552	0.388	0.466	0.504	0.605	0.1294	0.1079
0.441	0.444	0.445	0.485	0.567	0.592	0.437	0.524	0.484	0.581	0.1336	0.1113
0.556	0.559	0.560	0.612	0.697	0.723	0.450	0.540	0.517	0.621	0.1477	0.1147
0.702	0.704	0.706	0.773	0.866	0.895	0.462	0.562	0.534	0.641	0.1418	0.1182
0.885	0.887	0.888	0.974	1.08	1.12	0.473	0.568	0.547	0.656	0.1460	0.1216
1.12	1.12	1.12	1.23	1.34	1.38	0.483	0.580	0.554	0.665	0.1500	0.1250
1.41	1.41	1.41	1.55	1.66	1.69	0.493	0.592	0.554	0.665	0.1542	0.1285
1.41	1.41	1.41	1.55	1.62	1.65	0.493	0.592	0.535	0.642	0.1532	0.1276
1.78	1.78	1.78	1.95	2.04	2.07	0.503	0.604	0.551	0.661	0.1583	0.1320
2.24	2.54	2.24	2.47	2.54	2.57	0.514	0.511	0.549	0.659	0.1627	0.1355
2.24	2.24	2.24	2.47	2.53	2.55	0.515	0.518	0.545	0.655	0.1615	0.1346
2.82	2.82	2.82	3.10	3.16	3.18	0.525	0.530	0.557	0.665	0.1666	0.1388
3.56	3.56	3.56	3.92	3.97	3.98	0.536	0.543	0.561	0.673	0.1708	0.1423

Table A-4. *Inductive Reactance Spacing Factor (x_d) Ohms per Mile, per Conductor*

50 Hertz

SEPARATION — Inches

Feet	0	1	2	3	4	5	6	7	8	9	10	11
0	—	−0.2513	−0.1812	−0.1402	−0.1111	−0.0885	−0.0701	−0.0545	−0.0410	−0.0291	−0.0184	−0.0088
1	0	0.0081	0.0156	0.0226	0.0291	0.0352	0.0410	0.0465	0.0517	0.0566	0.0613	0.0658
2	0.0701	0.0742	0.0782	0.0820	0.0857	0.0892	0.0927	0.0960	0.0992	0.1023	0.1053	0.1082
3	0.1111	0.1139	0.1166	0.1192	0.1217	0.1242	0.1267	0.1291	0.1314	0.1337	0.1359	0.1380
4	0.1402	0.1423	0.1443	0.1463	0.1483	0.1502	0.1521	0.1539	0.1558	0.1576	0.1593	0.1610
5	0.1627	0.1644	0.1661	0.1667	0.1693	0.1708	0.1724	0.1739	0.1754	0.1769	0.1783	0.1798
6	0.1812	0.1826	0.1839	0.1853	0.1866	0.1880	0.1893	0.1906	0.1918	0.1931	0.1943	0.1956
7	0.1968	0.1980	0.1991	0.2003	0.2015	0.2026	0.2037	0.2049	0.2060	0.2071	0.2081	0.2092
8	0.2103											
9	0.2222											
10	0.2328											
11	0.2425											
12	0.2513											
13	0.2594											
14	0.2669											
15	0.2738											
16	0.2804											
17	0.2865											
18	0.2923											
19	0.2977											
20	0.3029											
21	0.3079											
22	0.3126											
23	0.3170											
24	0.3214											
25	0.3255											
26	0.3294											
27	0.3333											
28	0.3369											
29	0.3405											
30	0.3439											
31	0.3472											
32	0.3504											
33	0.3536											
34	0.3566											
35	0.3595											
36	0.3624											
37	0.3651											
38	0.3678											
39	0.3704											
40	0.3730											
41	0.3755											
42	0.3779											
43	0.3803											
44	0.3826											
45	0.3849											
46	0.3871											
47	0.3893											
48	0.3914											
49	0.3935											

x_d at
50 cycles
$x_d = 0.2328 \log_{10} d$
d = separation, feet.

60 Hertz

SEPARATION — Inches

Feet	0	1	2	3	4	5	6	7	8	9	10	11
0	—	−0.3015	−0.2174	−0.1682	−0.1333	−0.1062	−0.0841	−0.0654	−0.0492	−0.0349	−0.0221	−0.0106
1	0	0.0097	0.0187	0.0271	0.0349	0.0423	0.0492	0.0558	0.0620	0.0679	0.0735	0.0789
2	0.0841	0.0891	0.0938	0.0984	0.1028	0.1071	0.1112	0.1152	0.1190	0.1227	0.1264	0.1299
3	0.1333	0.1366	0.1399	0.1430	0.1461	0.1491	0.1520	0.1549	0.1577	0.1604	0.1631	0.1657
4	0.1682	0.1707	0.1732	0.1756	0.1770	0.1802	0.1825	0.1847	0.1869	0.1891	0.1912	0.1933
5	0.1953	0.1973	0.1993	0.2012	0.2031	0.2050	0.2069	0.2087	0.2105	0.2123	0.2140	0.2157
6	0.2174	0.2191	0.2207	0.2224	0.2240	0.2256	0.2271	0.2287	0.2302	0.2317	0.2332	0.2347
7	0.2361	0.2376	0.2390	0.2404	0.2418	0.2431	0.2445	0.2458	0.2472	0.2485	0.2498	0.2511
8	0.2523											
9	0.2666											
10	0.2794											
11	0.2910											
12	0.3015											
13	0.3112											
14	0.3202											
15	0.3286											
16	0.3364											
17	0.3438											
18	0.3507											
19	0.3573											
20	0.3635											
21	0.3694											
22	0.3751											
23	0.3805											
24	0.3856											
25	0.3906											
26	0.3953											
27	0.3999											
28	0.4043											
29	0.4086											
30	0.4127											
31	0.4167											
32	0.4205											
33	0.4243											
34	0.4279											
35	0.4314											
36	0.4348											
37	0.4382											
38	0.4414											
39	0.4445											
40	0.4476											
41	0.4506											
42	0.4535											
43	0.4564											
44	0.4592											
45	0.4619											
46	0.4646											
47	0.4672											
48	0.4697											
49	0.4722											

x_d at
60 cycles
$x_d = 0.2794 \log_{10} d$
d = separation, feet.

Table A-5. *Shunt CAPACITIVE Reactance Spacing Factor* (x_d) *Megohm Miles, per Conductor*

50 Hertz

	SEPARATION											
	Inches											
Feet	0	1	2	3	4	5	6	7	8	9	10	11
0	—	−0.0885	−0.0638	−0.0494	−0.0391	−0.0312	−0.0247	−0.0192	−0.0144	−0.0102	−0.0065	−0.0031
1	0	0.0028	0.0055	0.0079	0.0102	0.0124	0.0144	0.0164	0.0182	0.0199	0.0216	0.0232
2	0.0247	0.0261	0.0275	0.0289	0.0302	0.0314	0.0326	0.0338	0.0349	0.0360	0.0371	0.0381
3	0.0391	0.0401	0.0410	0.0420	0.0429	0.0437	0.0446	0.0454	0.0463	0.0471	0.0478	0.0486
4	0.0494	0.0501	0.0508	0.0515	0.0522	0.0529	0.0535	0.0542	0.0548	0.0555	0.0561	0.0567
5	0.0573	0.0579	0.0585	0.0590	0.0596	0.0601	0.0607	0.0612	0.0618	0.0623	0.0628	0.0633
6	0.0638	0.0643	0.0648	0.0657	0.0662	0.0666	0.0666	0.0671	0.0675	0.0680	0.0684	0.0689
7	0.0693	0.0697	0.0701	0.0705	0.0709	0.0713	0.0717	0.0721	0.0725	0.0729	0.0733	0.0737
8	0.0740											
9	0.0782											
10	0.0820											
11	0.0854											
12	0.0885											
13	0.0913											
14	0.0940											
15	0.0964											
16	0.0987											
17	0.1009											
18	0.1029											
19	0.1048											
20	0.1067											
21	0.1084											
22	0.1100											
23	0.1116											
24	0.1131											
25	0.1146											
26	0.1160											
27	0.1173											
28	0.1186											
29	0.1199											
30	0.1211											
31	0.1223											
32	0.1234											
33	0.1245											
34	0.1255											
35	0.1266											
36	0.1276											
37	0.1286											
38	0.1295											
39	0.1304											
40	0.1313											
41	0.1322											
42	0.1331											
43	0.1339											
44	0.1347											
45	0.1355											
46	0.1363											
47	0.1371											
48	0.1378											
49	0.1386											

x_d' at
50 cycles
$x_d' = 0.08198 \log_{10} d$
d = separation, feet.

60 Hertz

	SEPARATION											
	Inches											
Feet	0	1	2	3	4	5	6	7	8	9	10	11
0	—	−0.0737	−0.0532	−0.0411	−0.0326	−0.0326	−0.0206	−0.0160	−0.0120	−0.0085	−0.0054	−0.0026
1	0	0.0024	0.0046	0.0066	0.0085	0.0103	0.0120	0.0136	0.0152	0.0166	0.0180	0.0193
2	0.0206	0.0218	0.0299	0.0241	0.0251	0.0262	0.0272	0.0282	0.0291	0.0300	0.0309	0.0318
3	0.0326	0.0334	0.0342	0.0350	0.0357	0.0365	0.0372	0.0379	0.0385	0.0392	0.0399	0.0405
4	0.0411	0.0417	0.0423	0.0429	0.0435	0.0441	0.0446	0.0452	0.0457	0.0462	0.0467	0.0473
5	0.0478	0.0482	0.0487	0.0492	0.0497	0.0501	0.0506	0.0510	0.0515	0.0519	0.0523	0.0527
6	0.0532	0.0532	0.0540	0.0544	0.0548	0.0552	0.0555	0.0559	0.0563	0.0567	0.0570	0.0574
7	0.0577	0.0581	0.0584	0.0588	0.0591	0.0594	0.0598	0.0601	0.0604	0.0608	0.0611	0.0614
8	0.0617											
9	0.0652											
10	0.0683											
11	0.0711											
12	0.0736											
13	0.0761											
14	0.0783											
15	0.0803											
16	0.0823											
17	0.0841											
18	0.0858											
19	0.0874											
20	0.0889											
21	0.0903											
22	0.0917											
23	0.0930											
24	0.0943											
25	0.0955											
26	0.0967											
27	0.0978											
28	0.0989											
29	0.0999											
30	0.1009											
31	0.1019											
32	0.1028											
33	0.1037											
34	0.1046											
35	0.1055											
36	0.1063											
37	0.1071											
38	0.1079											
39	0.1087											
40	0.1094											
41	0.1102											
42	0.1109											
43	0.1116											
44	0.1123											
45	0.1129											
46	0.1136											
47	0.1142											
48	0.1149											
49	0.1155											

x_d' at
60 cycles
$x_d' = 0.06831 \log_{10} d_1$
d = separation, feet.

Table A-6. *Circuit Constants of Some Typical Cables (per conductor)*

Voltage Class kV	Construction	Wire Size	Resistance ohms per Mile	Inductive Reactance ohms per Mile	Capacitive Reactance ohm-Miles
1	3 conductor Belted, paper insulation	4	1.58	0.175	5400
15	3 conductor Belted, paper insulation	0000	0.310	0.174	5600
69	Single conductor Paper insulated Oil filled, 6 inch spacing	250,000 cir. mil	0.263	0.334	4790
230	Single conductor Paper insulated Oil filled, 6 inch spacing	1,000,000 cir. mil	0.067	0.276	7140
525	Single conductor Paper insulated Pressurized, oil filled	500,000 cir. mil	0.113	0.25	7600
525	Single conductor Gas insulated Pipe type (Experimental)		0.0086	0.172	32,000

Table A-7. *Typical Constants of Three-Phase Synchronous Machines*

Reactances are per unit, values below the line give the normal range of values, while those above give an average value.

	1	3	4
	x_d (unsat)	x_d' rated voltage	x_d'' rated voltage
2-Pole turbine generators	$\dfrac{1.20}{0.95\text{–}1.45}$	$\dfrac{0.15}{0.12\text{–}0.21}$	$\dfrac{0.09}{0.07\text{–}0.14}$
4-Pole turbine generators	$\dfrac{1.20}{1.00\text{–}1.45}$	$\dfrac{0.23}{0.20\text{–}0.28}$	$\dfrac{0.14}{0.12\text{–}0.17}$
Salient-pole generators and motors (with dampers)	$\dfrac{1.25}{0.60\text{–}1.50}$	$\dfrac{0.30}{0.20\text{–}0.50(\dagger)}$	$\dfrac{0.20}{0.13\text{–}0.32(\dagger)}$
Salient-pole generators (without dampers)	$\dfrac{1.25}{0.60\text{–}1.50}$	$\dfrac{0.30}{0.20\text{–}0.50(\dagger)}$	$\dfrac{0.30}{0.20\text{–}0.50(\dagger)}$
Condensers, air cooled	$\dfrac{1.85}{1.25\text{–}2.20}$	$\dfrac{0.40}{0.30\text{–}0.50}$	$\dfrac{0.27}{0.19\text{–}0.30}$
Condensers, hydrogen cooled	$\dfrac{2.20}{1.50\text{–}2.65}$	$\dfrac{0.48}{0.36\text{–}0.60}$	$\dfrac{0.32}{0.23\text{–}0.36}$

(†) High speed units tend to have low reactance and low speed units high reactance.

Table A-8. *Transformer Reactance, Percent*

Distribution transformers	3 to 500 kVA
kV	%
2.5	1.2–5.0
15	1.2–5.0
25	4.0–5.5
69	6.0–6.5

Station transformers	500–5000 kVA
kV	%
2.5	5.5
15	5.5
34.5	6.0– 6.5
69	7.0– 7.5
138	8.0–10.5
230	9.0–11.0

Station Transformers	Above 30000 kVA	
kV	%	
	A*	B*
34.5	5.0– 8.5	8.7–14
69	6.0–10	10 –16
138	6.6–13	10 –20
230	7.5–16	13 –27
345	8.5–18	14 –30
500	9 –21	15 –34
725	11 –22	23 –35

*A—Oil-immersed self-cooled, and forced air cooled.
*B—Oil immersed, forced oil cooled with water cooler.

The higher values of impedance are associated with the higher kVA rating in the voltage class.

Transformer resistance is usually well below 1 percent.

Index